Also By Richard L. Handy

The Day the House Fell (non-technical)

Soil Engineering (co-author) 3rd edition

Soil Engineering (co-author) 4th Edition

Geotechnical Engineering: *Soil and Foundation Principles and Practice, 5th Ed.*

The Papers of R. L. Handy, edited by D. J. White and A. J. Lutenegger.
Samples:

The Arch in Soil Arching (ASCE Middlebrooks Award)

Loess Distribution by Variable Winds

The igloo and Natural Bridge as Ultimate Structures

Alluvial Cutoff Dating from Subsequent Growth of a Meander

First-Order Rate Equations in Geotechnical Engineering

FORE AND THE FUTURE OF PRACTICALLY EVERYTHING

A Scientific Look into the Future

(or a guess with a college education)

Richard L. Handy Ph.D.

Moonshine Cove Publishing, LLC
Abbeville, South Carolina U.S.A.

ISBN: 978-1-937327-31-6
Library of Congress Control Number: 2013951585

Praise for Richard L. Handy and Fore and the Future

"An utterly fascinating book built around an astonishing variety of day-to-day practical applications of what is known as a mathematical First Order Rate Equation. Along the way the reader is presented with a cornucopia of factual information that is educational all by itself."—*Gordon Eaton, Ph.D., former President of Iowa State University and Director of the U.S. Geological Survey.*

"...I hope it hits a home run."—*Alan Alda, actor and science spokesperson .*

"This witty, very readable book employs amusing anecdotes and analogies to explain how a seemingly simplistic mathematical concept can describe a wide range of observable phenomena, from coffee sweetness to climate change. It is a must read for anyone curious about how and why things work the way they do...a great book, which hopefully will get the wide readership it deserves."—*Michael Mann, Ph. D., Distinguished Professor and Director, Earth System Science Center (ESSC), Pennsylvania State University.*

"A perfect example [of writing humor without losing the continuity of the primary message] is a new book by Emeritus Professor Richard Handy. The book addresses the complex subject of rising sea levels and its impact on Planet Earth. Dr. Handy makes the subject easily understood and fun to read – primarily due to his skillful use of appropriate humor.

"*FORE and the Future of Practically Everything* is the title of the new book. It is published by Moonshine Cove Publishing." —*Jim Woodard, columnist and story teller at the Ronald Reagan Library*

About the cover: A skipping stone moves with progressively shorter steps until it slows down and sinks.

That is the world according to FORE, which stands for "First-Order Rate Equation."

Photo courtesy of the Author and the Stone Skippers.

Dedication

To my lovely wife Kathryn,
who tried to warn me and I wouldn't listen.

The views and opinions expressed in this book are solely those of the author and nobody else in particular.

Acknowledgments

The late Prof. T. Demirel guided the author into the lure of "first-order rate equations," which in this book are modified and given an acronym "FORE." Suggestions and encouragement from many former students, friends and associates including Dr. Nat Fox have given valuable insights and encouragement

The author's own mentors include Dr. Leonard Feinberg, whose specialty was humor and who saw a potential in FORE. He also held that if it's funny write it down and let it out.

Finally, the author thanks the many former students and associates including those who recently endowed a Professorship in his name at Iowa State University, and to the many others who were considering it and then thought better of it.

Foreword

The principal mentor in my life, Dick Handy, in this brilliant and humorous book explains how he may have inherited his grandfather's humor genes. I believe that he did just that. I also believe that he must have simultaneously inherited some pretty decent scientist-engineer genes. This is a science book with a pervasive sense of humor! Always the teacher, Handy takes a reader (even one with limited mathematical skills), on a trip through equations and then into logarithms! It includes the first popular definition of logarithms as a chorus of singing lumberjacks! FORE looks to the past to project into the future. As Dick says, it is "*a guess with a college education.*"

FORE stands for "*First-Order Rate Equation*" that the author modified so it can apply to a broad spectrum of physical processes - from life expectancy, to home run hitting, to the rise in sea level. The FORE approach is not a prediction, it is a *projection* that uses a rather strict equation, and is based on real data. This book is magnificently entertaining while being extremely enlightening. Likely readers will range from inquiring high school students to prominent scientists and engineers.

I firmly believe that this book will generate at least one career-changing breakthrough. This breakthrough may emerge as a patented process - or perhaps as a crucial decision derived from accurately forecasting the future. Regardless, I believe that at least one reader will become wealthy, or famous, or both, as a direct result of reading and absorbing this amazing book. FORE does, after all, open the possibility to look into the future before getting there.

As Dick Handy says, "*Remember, you saw it here first.*"

Nat (Nathaniel) S. Fox, Ph.D., P.E.
Inventor, Geopier and Impact Pier foundation systems
Graduate of West Point and Iowa State University
Sustained by wife, Lily Kaye Fox

CONTENTS

FORE and the Future of Practically Everything
A guess with a college education.
R. L. Handy, Ph.D.

INTRODUCTION

When I was a boy my grandfather took me fishing along the banks of the Wapsipinicon River in Iowa. I learned that fishing is a time for quiet contemplation because talking might scare the fish. That probably contributed to a lifetime of reticence because in the back of my mind I always worried that I might scare the fish.

When the fish weren't biting, which was most of the time, I would lay down my fishing pole and throw a flat stone so it skipped over the water. I discovered that no matter how hard I threw the stone and despite my sincerity, the skips always made progressively shorter steps until finally the stone sank.

"There's a lesson in that," my grandpa said. I thought he was going to say something funny, or maybe a reminder about scaring the fish, but instead he said, "That's the nature of things—they start off fast and then they slow down and stop." Then he laid back, closed his eyes, and took a nap. That was my grandpa. He called it "truth by demonstration." When his disposition called for a cup of coffee and a doughnut, he had a cup of coffee and a doughnut. When it called for a nap, he took a nap. Truth by demonstration.

Years later my grandpa did one last truth by demonstration. For me it was a very sad time but as he said, that it was in the nature of things. He was on a hospital bed, the pain was under control, and he started telling stories. He and his brother had been orphaned as children and they must have used humor to cope. He told how one night at bedtime his older brother used a novel method to blow out the kerosene lamp, and the flame singed his butt, woke everybody up and almost set the house on fire. He told how they put a red-hot poker to the touch-hole of a muzzle-loader shotgun to shoot a prairie chicken and there was so much smoke they couldn't find the chicken. He told how on one Halloween some of the local boys took a teacher's Model T Ford apart and put it back together on the roof of her house, and how the next morning when she saw it she fell down laughing, gave them some cookies and then made them take it down.

We listened to stories and we laughed; my how we laughed! We laughed until there were tears in our eyes. Some nurses came running to see what was causing the commotion and they stopped, listened and they laughed. Soon the hospital room was nearly full of people listening to Grandpa's stories, laughing and crying at the same time. For the moment this was not just my grandpa; this was everybody's grandpa, reminiscing and telling funny stories. Then my grandpa paused and smiled and said he was ready to go to sleep now. He sighed and closed his eyes and went to sleep.

That is my last remembrance of my grandpa. Was this a reverie that can occur when the pain eases in one's final hours? Perhaps, but I think it was more than that. I think that my grandpa used humor to make life more fun, to help get over the hard spots. I like that idea. Maybe it's in my genes.

This book uses a special kind of telescope that looks back in order to look ahead. That is, it uses information from the past to try and get a handle on the future. This is not ordinary speculation and it is not fully ordained and consecrated speculation; it is *scientific* speculation. It is speculation that is based on a simple rule called a "first-order rate equation." Hence the acronym, *FORE*. The acronym also has a nice forward look about it.*

---

*The acronym was suggested by Prof. K. S. Li of the University of Hong Kong in a discussion of one of the author's papers. No kidding.

The rule. My grandfather liked sugar in his coffee, and observed that the sugar stopped dissolving as soon as there was no more sugar left to dissolve. He therefore would add more sugar, take a sip and say, "This coffee is too sweet!" We would respond, "Why don't you stop using so much sugar?" and he would look puzzled and say, "It wasn't too sweet to start with."

Adding more sugar still does not change a fundamental rule, that the sugar will dissolve slower as there is less left to dissolve. That is the story of FORE. It describes a process—any process—that slows down as it uses up a key ingredient. The concept comes from chemistry. This book is the first to try it out on a variety of physical processes.

An example is production from an oil well, which starts to slow down as soon as the well starts pumping. The production from entire oil fields starts to slow down as soon as they start pumping. At one time the U.S. was the world's leading oil producer, but no longer, and the same equation applies to all other oil fields including those in Saudi Arabia, the North Sea and Alaska. New oil discoveries make headlines, but again, as more oil is discovered, the less there will be remaining to be discovered. "Fracking" can get out more oil but that still does not create any more oil. The conditions look right for FORE, so maybe FORE can look into it.

Ups and downs. FORE can work for phenomena that are going up, like life expectancy, as well as those that are going down, like petroleum production. Either way the rates must be slowing down. Sport records go up, as in heights achieved in pole vaulting, or they can go down, as the time required to run the 100 meter dash. There must be physical limits, so what are those limits? How high can a person ever be able to jump; how fast can he or she ever run? In baseball, what is the maximum number of home runs that can be hit in a single season without a boost from doping? Can FORE also give an opinion?

Major events. Sometimes something comes along that resets the clock for FORE. In the case of the pole vault, it was introduction of the fiberglass pole. Old records can fade into oblivion as a result of an improvement in technology, which hardly seems fair. In the language of FORE that was a "major event," and FORE can salvage the old records by translating them into modern equivalents. How fast might Jesse Owens have run if he had worn Nikes and run on paved tracks, instead of running in kangaroo leather shoes on cinder tracks? Might he even still hold a world record? FORE can answer that.

FORE also can define the "gender factor" in sports records, to enable those set by women to be for fairly compared with those set by men. This will have increased relevance with an increase in transgender participants.

Limitations. FORE does not always work; it only works for processes that are slowing down. FORE does not work for a rate that is speeding up or staying the same. However, when it does work it can be used to project the past into the future. It is not "futurist;" it is "projectionist."

It is a guess with a college education.

Chapter 1. The Science of Slowing Down

Jack and Jill went up the Hill to have a little action
Jack fell down and broke his crown and spent a month in traction.

Men realize that there is an art to running out of gas. They run out of gas far more frequently than do women because men know in their hearts that all gas gauges exaggerate. They know that in order to be on the safe side a gas gauge will read EMPTY when there still are a couple of gallons left in the tank. Men therefore recognize that an unnecessary stop at a gas station is a waste of time. They know that they don't have to stop that often. They know that by a simple understanding of the inadequacies of the gas gauge they can start late and arrive early. That is why women don't run out of gas. A woman will keep an eye on the gas gauge and when it gets to the half-way mark she will start looking for a place to fill up.

Since running out of gas can be an inconvenience, it is good idea to make a practice run. A man can do this in order to evaluate automotive performance under less than ideal operating conditions. We performed the experiment with a pickup truck, wife driving, after a little sweet talk and a promise to go out for dinner. A pickup truck was used because pickup trucks are more likely to be driven by men and therefore are more likely to run out of gas.

How to simulate running out of gas. The main objective when running out of gas is to coast all of the way to a gas station, stop at the pump, then jump out and yell, "*Voila*!" while pretending that the achievement is nothing unusual. Coasting is accomplished by simply shifting into "Neutral" and allowing a vehicle to coast, but it should be emphasized that steering also is required. This kind of activity does not have legal sanction so it is best achieved on a country road with very little traffic and no deputy sheriff looking to render a legal opinion.

In our experiment we took the speed up to about 60 miles per hour on a level stretch of road, shifted into neutral, and coasted. Declining speeds were read from the speedometer into a portable tape recorder: 55...50...45... and so on. We coasted for what seemed like a very long time, until a farm tractor that looked like it was half the size of Minneapolis came along and ended the experiment. The total time was less than two minutes. Then it was time for dinner.

FORE as in FOREcast

Science is more than just opinion, particularly when it involves scientific measurements like how fast a pickup truck slows down coasting. FORE says that the rate of decline in speed should depend on the speed itself, that slowing down goes slower the more that something slows down. There is poetry in that, and if it works it should extend on past the actual measurements and into the future. FORE therefore does not make predictions; it makes *projections*. FORE is not futurist; it is projectionist. If something happens that changes the conditions, like running off the road or coming to a stop sign, that is called a "major event."

Keeping tally. After dinner, the tape was played back while using a running stopwatch to keep track of the time spent coasting. With every 5 mile-per-hour decrease in speed, the time on the stopwatch was read to the nearest second and written down using a ballpoint pen obtained from the post office, which shows how meaningful research can be accomplished without a lot of fancy equipment and a minimum of government funding. Great high school science project.

The numbers are in the following table. The experiment was performed only once so there was no cherry-picking to select the best data. At time zero on the stopwatch, the speed was 58 miles per hour; after 2 seconds it was 55 miles per hour; and so on. Important parts are circled.

Table 1. Data from coasting.

Speed (mph):	Y =58	55	50	45	40	35	30	25	20	15	10
Seconds:	X = 0	2	8	13	20	29	43	59	71	86	113

Oh, really? The circle at the left shows that slowing from 55 to 50 miles per hour only took *8 - 2 = 6* seconds. The dashed circle at the right shows that slowing down from 15 to 10 miles per hour took 113 - 86 = 27 seconds. That is over four times as long to slow down by the same amount! As anticipated, the slower we were going, the slower we slowed down. Coasting therefore has a *master plan*.

Establishing a relationship. One way to establish a relationship is to find a quiet place and park, but that's another story. The relationship between numbers in the table can be demonstrated by drawing a graph. In Fig. 1 the time measurements were stepped off to the right starting with zero at the left. That is called the "x direction." Each corresponding speed was stepped off upward starting with zero at the bottom, *up* being the "y direction." The data points line up in a smooth curve that looks like the path of an airplane coming in for a landing.

A computer has such a love relationship with numbers that it can describe a graph with something incomprehensible called an *equation*. The equation in Fig.

1 is thoroughly obnoxious. It is an *empirical equation,* and shows how empires are overblown and lack character. Empirical relationships change with the evening news, particularly if they deal with health issues.

Analytical relationships. Another approach is to first think things over and then test to see if data fit that way of thinking. That is an *analytical* approach. It depends on logic so it has to be kept simple. The equation in Fig. 1 is not analytical and not simple, and makes about as much sense as a board meeting after a three-martini lunch. The equation is written in a universal language called "algebra," but we shouldn't hold that against it.

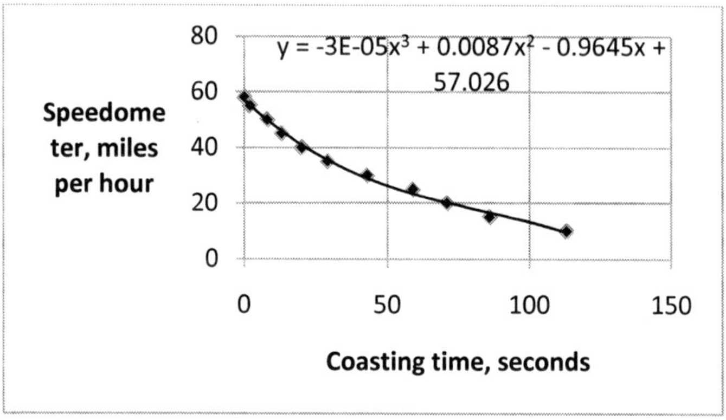

Fig. 1. Graphs make data visual, but without some spiritual guidance the equation can get quite nasty.

Having trouble with algebra. A first contact with algebra usually is when older kids tell younger kids, "You think LONG DIVISION is hard; wait until you take ALGEBRA!" The older kid then slaps a hand to the jaw, rolls his eyes, falls down and looks silly. The scare tactic is not good, so what is being done about it? Easy; quit teaching long division, but then there is no warning.

Algebra was invented by a ninth-century sheik named Al Gebra after his camel stopped at a railroad crossing and he couldn't remember his phone number so he wrote "X" on the camel's hump to serve as a reminder. He then made his most famous observation, that "X marks the spot," after which the camel marked the spot. Actually it takes both an "X" and a "Y" to mark a spot, as the spot where a camel trail crosses a railroad track.

A simple equation. The equation for every straight line in the entire universe is as easy as chicken broth:

$$y = ax + b \qquad (1)$$

where x and y represent critical parts of the chicken and a and b are the water and the salt. In the equation "ax" does not have its usual meaning, which is lucky for the chicken. This is not English; it is algebra, and "ax" means that a number "a" is multiplied times a number "x." The equation says that a number "y" equals a times x plus another number, "b." A graph of this relationship makes a straight line. The values of "a" and "b" define the line while the "x" and "y" are floaters. A different straight line only has different values for a and for b. That is nice.

An easy example is if $a = 1$ and $b = 0$. If those numbers are put in for a and b in the equation, $y = x$, and that is about as easy as it can get. Then if $x = 4$, $y = 4$, if $x = 19$, $y = 19$, and so on. That graph is a straight line that starts at the lower left corner and goes up at a 45 degree angle. This is the *only* line for the *only* relationship that does that with those particular values of a and b.

Logarithms

There is one more hitch: FORE uses logarithms, so WHAT in Holy Green Hell are logarithms?

The Logarithms (capital L) achieved fame as a choral group of lumberjacks who performed Rock and Roll where it really got started, in tops of swaying pine trees. The group used chain saws instead of Gibsons because chain saws do not require any amplification. Their big hit was "Thunder and Crash, Little Nellie," which they concluded by revving up their chain saws and cutting down a tree. Between sets the camp cook would dance around naked and play *Blue Moon* on his ukulele until one cold day when he strummed his personals into the strings and started yipping like a coyote caught in a jaw trap and the high tenor laughed so hard that he fell out of his tree, which would not have been so bad except that the baritone tried to catch him by reaching out with his chain saw, which happened to be running at the time. OSHA therefore issued a directive requiring safety shields on all ukuleles, and the cook was so disgusted that he quit the ukulele and took up the clarinet. At the mere mention of OSHA he would stomp his feet and make it chatter like a squirrel.

The Logarithms eventually had a falling out but on a quiet day a clarinet still can be heard playing somewhere off in the distance although some people who lack sensitivity claim it's only Bigfoot with asthma, or only tree branches rubbing in the wind.

Logging down the national debt. Now let's get serious: Here is how the National Debt can be brought under control by a simple application of logarithms. The logarithm of 10 is 1, the logarithm of 100 is 2, of 1000 is 3, and so on. That means that every time the national debt gets multiplied by 10, *all that is required is to add another zero, and zeros don't cost anything!* The debt can be in billions, trillions, or septaquadrillions, and it matters *not* what your country can do for you, but what you can do for your country by a proper application of logarithms. All that will be required is to write "Logarithm" on the tax form. Owe a $1000? No, indeed! The logarithm is 3.

Logarithm box score. A more formal definition of a logarithm is a bit clumsy: It is how many 10's are multiplied together to get a certain number. The logarithm of 100 is 2 because 10 x 10 = 100. For 1000 it is 3. So what about a number like 50? The logarithm is not a whole number; it is between 1 and 2, in this case 1.7.

"Natural logarithms" are organic, so naturally they cost more. The natural logarithm of 10 is not 1; it is 2.3. Natural logarithms use "ln" instead of "log" because "log" reminds people of the redwoods. Natural logs are the language of mathematicians, Socialists and girls who ride unicycles. As a courtesy we discuss natural logarithms in the back closet of this book called an Appendix where they will be less annoying.

FORE and logarithms. Logarithms create a HUGE shining advantage for FORE because they can change a curve such as in Fig. 1 into an equivalent straight line in Fig. 2. Logarithms turn Swan Lake into Moon Walk. The graph in Fig. 2 is exactly the same as in Fig. 1 except that the speed scale is converted to logarithms. The scale is stretched out for lower speeds and compressed for higher ones, and it is done systematically. The graph is only "semilogarithmic" because only the speed scale is converted. If time is converted, the line will not be straight and the entire concept would be out the window. Some older professors are on record as having struggled with this, so we have to keep an eye out and try and be charitable.

In the FORE equation, y values are expressed as logarithms. The equation changes to:

$$\log y = ax + b \qquad (2)$$

The straight line in Fig. 2 is represented by one and only one equation. It is:

$$\log y = -0.0065x + 1.7509$$

The *a* and *b* numbers look a little ragged but that is because we have come a long way from counting on fingers. The *a* is negative. That means that the line slopes down because the truck was slowing down.

Even though the graph in Fig. 2 uses the same data as the graph in Fig. 1, the equation is so much simpler and more manageable! Not only that, if the experiment were to be repeated regardless of whether it is with a Ford Escort or Lamborghini, only the values of *a* and *b* would change. Otherwise the equation will remain exactly the same.

Testing a relationship

A test for FORE describes how well data fit a straight line. The test is called "R^2" (read "R-squared"), which is available on a computer spreadsheet. R^2 is calculated from the *x* and *y* number pairs. If it is exactly 1.0, that means perfect, with every data point exactly on the line. It also can mean there are only two data points so it is impossible for them not to be in a straight line. If R^2 is exactly 1.0 it also can mean that somebody cheated because measurements in a continuum are not perfect. The R^2 value for the line in Fig. 2 is *0.9957*, which is high enough to be convincing and low enough to show that nobody cheated. The time measurements were not exact but were to the nearest second, which contributes to "experimental error." Experimental error also is called family planning but nobody names their kid that.

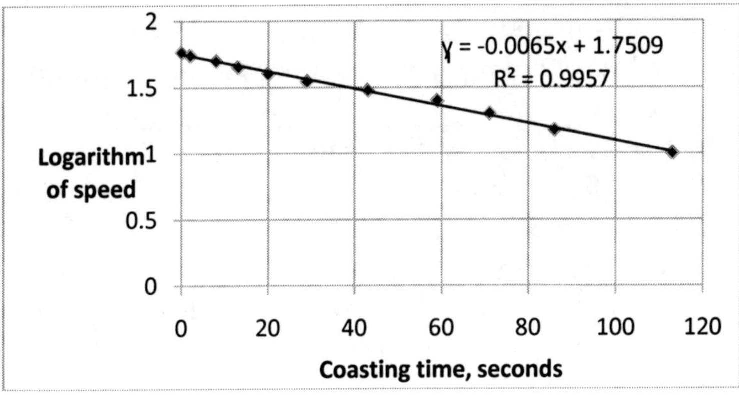

Fig. 2. Data from Fig. 1 re-plotted with speeds converted to logarithms. The equation is brief and the straight line means that FORE worked. The line is not perfect but neither were the road or the people performing the test.

FOREcast

The coasting experiment lasted all of 2 minutes but the equation for a straight line has no limits. What if the experiment had been continued for 5 minutes; what would be the speed then? The equation can project into the future; put 300 seconds in for x , use the equation to determine y, and reverse the process that changes a number into its logarithm. The answer comes out to be less than a mile per hour so the pickup might have been stopped if it hit a grasshopper, but it would have to be a significant grasshopper.

Why use FORE? Why not use the equation in Fig. 1 to project into the future? Its R^2 is even higher, 0.9995 but it has no guidance. It says that after 5 minutes the speed will be: *-259 miles per hour!* Not only is that many times too fast, the minus sign means *backing up!* Score one for FORE. The lesson is that an empirical relationship can't be pushed *where it doesn't already go*. FORE can do that because it is not empirical; it is *analytical.* Slowing down means a loss of energy, which happens as one gets older.

Box score. Let's say that the experiment had been continued for 5 minutes; what would be the speed? Plugging 300 seconds in for *x* in the equation gives:

$\log y = -0.0065(300) + 1.7509,$

where ( ) means to multiply. Multiplying gives -1.95, so

$\log y = -1.95 + 1.7509$, which with a little arithmetic gives

$\log y = -0.209$

Now we "antilog" the logarithm, which is 10^x on the keyboard of a calculator that does logarithms. That gives y = *0.617 miles per hour,* about as fast as fog lifting.

FORE, a special case. What if a Ford Escort is coasting down a very long hill so it slows down but does not come to a stop; then how does FORE work? That is in the next chapter, using data from the world's largest icebreaker cruising the fabled Northwest Passage, a multimillion dollar expedition that ended up using exactly the same tools, a stopwatch and a tape recorder.

Important answers to important questions:

1. *Does FORE predict the future?* NO! FORE does not make *predictions*; it makes *projections.* FORE definitely is not "futurist;" it is "projectionist."

2. *Is FORE "singularity?"* NO! Singularity describes a rate that is increasing, like a rocket that is taking off. FORE can take over when the rocket runs out of fuel.

3. *Is FORE a* LAW? A scientific law is not the same as a legal law, for which we should be eternally grateful. A scientific law cannot be lobbied, argued, voted, passed, or

vetoed, nor can it be ignored by thieves, murderers, drunks, prostitutes, politicians, and hedge fund managers. A scientific law only states facts. Its language is algebra, as in "$e = mc^2$." A scientific law is a given and does not include an explanation. An explanation is a human activity and is called a *hypothesis.*

4. *What is the difference between law, theory and hypothesis?* A scientific law is a statement of fact *that is supported by scientific data*. It is not opinion. Opinion is only ignorance disguised by insistence.

A hypothesis is an explanation, and there can be more than one possible explanation. Different hypotheses therefore are devised, compared and tested.

A hypothesis that passes all tests can graduate into a *theory*, which in science is as far as it can go. Darwin observed evolution and hypothesized "survival of the fittest." His hypothesis then graduated into a theory. It means that "just a theory" is not a criticism; it is an endorsement.

Fig. 3. Law of the Skipping Stone; it slows down and sinks. FORE is a hypothesis to explain the Law.

Chapter 2. On Breaking the Ice

The S. S. Manhattan went aground and was cut up for scrap, which was a sorry end for the bidets, and certainly not the end that they had been accustomed to.

In 1967 a large amount of crude oil was discovered under the northern rim of Alaska, but it hardly made a dent in the local blubber market. The oil was not worth anything unless a way could be found to get it out. The cool way would be by ship across the roof of Canada through the fabled Northwest Passage, but at the time the Passage was perennially clogged with ice. That meant that an oil tanker also would have to be an icebreaker.

At that time the largest oil tanker in the world was the S. S. Manhattan. It was surgically enhanced by welding boiler plate along the sides, and it was given a nose job that stuck out like an inverted beak. That hopefully turned it into the world's largest-ever icebreaker. The ship then embarked on the most extensive and expensive full-scale study of ice breaking ever attempted. Meanwhile just to be safe negotiations were under way to build a pipeline.

A very Long wall

For a long time the Americas were looked on as only a wall that ships had to go around to get to China. The only path was through the Strait of Magellan, a channel about 350 miles (570 km) long separating Antarctica from the southern tip of South America. At its narrowest it is only a little over a mile (2 km) wide, and pinches the circumpolar west-to-east sea current so it holds ships back like giant hand. Being close to the huge polar ice mass, winds are strong and the weather nasty. Masts of sailing ships sometimes would light up with St. Elmo's fire to add more awe and mystery to the icy conditions. There had to be a better way, perhaps a Northwest Passage across the roof of Canada.

Henry Hudson sailed into Hudson Bay in the 1600s and thought it must be the Northwest Passage but it wasn't. Then things did not go well for him, particularly after the mutiny. In the mid-1800s Sir John Franklin was a natural choice to lead an expedition because of his prior experience in the Arctic when he became known as "the man who ate his boots." It probably inspired a famous scene in Charlie Chaplin's *The Gold Rush,* in which he boiled his boot, carefully dissected it with his knife and fork, and sucked the juice out of every nail.

At the time Franklin was the Governor of Tasmania, located down under and off to one side. His expedition embarked from England in 1845 with two sailing ships, the *Erebus,* which signifies darkness, and the *Terror,* which needs no further explanation. Both ships had auxiliary steam engines that put out 25 horsepower, about the same as for today's small fishing boats.

When the Franklin expedition was about halfway through the Passage, both ships got stuck in the ice. After three years sitting around waiting for the ice to melt, Franklin was dead and the crews were getting moody. Their food was preserved by a new process called "canning" in which tin cans were sealed with lead solder, which may explain some of the decision-making.

Maritime regulations require lowering the lifeboats before abandoning ship, which makes sense as long as there is a place to put them. The lifeboats were lowered onto the ice, piled full of non-essentials and then laboriously dragged south in a search for open water.

Eskimos who saw the procession did not like the look of it, partly because there was no way that a group that large could survive by hunting, even if they knew how to hunt. The only traces that ever were found were some silver place settings and human bones that were suspiciously sawed to the diameter of a cooking pot. The last man out probably choked on an oar.

First time through. Sixty years later, in 1903-1906, an unflappable Norwegian named Roald Amundsen beefed up a 70-foot (21 m) wood herring boat named the Gjøa and went all of the way through the Passage. It took him two years so he relaxed, made friends with the Inuit (Eskimos), and learned their survival methods. Five years later he dressed in furs and used dog sleds to be the first to reach the South Pole. Another expedition led by Sir Robert Scott regarded it as more sporting if they would pull the sleds themselves. They also got to the Pole, but on the way back ended up getting stiff upper lips the hard way. Dog sleds still have an advantage for Arctic winter travel because dogs can divert the attentions of polar bears, and in an emergency it is not easy to eat a snowmobile.

A love boat

The Manhattan was the pleasure oil tanker of Greek shipping magnate Aristotle Onassis. The ship bridge, officer and guest quarters were located amidships away from engine noise. Some of the rooms had snatches of Italian art left over from a fling with an opera diva until Jackie Kennedy came along and rocked the boat. The Manhattan also may have been the only oil tanker ever to be outfitted with feminine cultural devices called bidets. Years later it went aground during a typhoon and was cut up for scrap, which was a sorry end for the bidets and certainly not the end that they had been accustomed to.

The first run by the Manhattan through the Northwest Passage got all of the publicity. It was in 1969 when the ice was thinnest, and the ship made it through. A second run was attempted in 1970 to get more data, and was kept quiet. The first trip did not have enough ice; the second had too much.

Pressure ridges. Floating pack ice cracks open and moves in the wind to create open "leads" that are inviting to ships, then clamp shut when the wind changes. The pack ice exerts so much pressure that it crumples itself up into "pressure ridges," and a boat caught up in the pressure can get crumpled up likewise. Amundsen's boat was small and designed to "pop up" and sit on top of the ice. The Manhattan was over 1000 feet long and did not do that. Instead steel plates were welded along the sides to deflect the ice pressure downward, and the new prow was designed to push down and break the ice. A Canadian icebreaker went along just in case.

The objective of the research was to determine how much power would be required for a large ship to smash through different thicknesses of polar ice. The Manhattan did not have enough power.

Ram-a-damn

When an icebreaker gets stuck, the procedure involves backing up and ramming ahead to have another go at it, like a hammer driving a nail. Sometimes if the ice is thin, the ship may keep going but at a lower speed. After a few decades to think about it, this looked like it might be an opportunity for FORE, like coasting but not to a stop. That throws another unknown quantity, the final speed, so it will be interesting to see if the equation still works.

Fig. 5 shows the record from a single ram into the ice, where the impact speed was 12 knots and the final speed was around 4 knots. After a while the Captain mumbled something about going too damn slow and had the propellers reversed to back the ship up for another ram.

A problem. Since the final equilibrium speed is not zero, it has to be guessed at so it is appropriate to have a clue that can help to determine if the guess is right. Instead of the final speed being only "y;" it is "$y - c$" where c represents the unknown final speed. Equation (2) gives way to

$$log(y - c) = ax + b \qquad (3)$$

That is how it is in science; when an equation doesn't work, fix it. However, in this case if c is not known, the equation not solvable. It is like handing the kids the car keys and saying you will meet them somewhere. It means that more information is required, but fortunately FORE gives a clue; if the equation is right, it will make a straight line. Then all that is necessary is to try different values for c and find one that makes a straight line. This approach is called "trial and error." It is like dating but instead of beer and pizza it uses a computer.

Trying trial-and-error. Fig. 4 is a FORE graph resulting from a particular ramming experiment, with three lines for three different trial values of c, and the middle gives the best straight line. It is for $c = 4.3$ *knots,* which would have been the final speed if the ram had continued to play out under the same conditions.

This happens to be HUGE, which means really big. It is big because the same approach can be used for a lot of things, everything from world records in sports to food production. It opens a possibility to look into the future of all those things before getting there. This has to be better than some fanciful prediction because it is a *projection* that uses a rather strict equation and is based on real data. Remember, you saw it here first.

A nautical knot does not relate to bondage but is a nautical mile per hour. A land mile originally represented 1000 paces of a Roman soldier with long legs, and a nautical mile is defined on the basis of the distance around the equator. It equals 1.15 miles (1.85 km).

Testing FORE

FORE can be tested using statistics, which should not to be confused with sadistics even though they can create the same impression. A statistic that can be used for FORE is called "R^2" (pronounced R-squared).

If $R^2 = 1.0$ it means that a straight line is perfect, that all of the data points are exactly on the same straight line. In Fig. 4 the best straight line gives the highest $R^2 = 0.998$, which is really good, which is how it should be with an analytical approach. The other approach is called "empirical," and empires can crumble. It an attempt to relate a bunch of measurements without a plan. For example, it can be used to evaluate the effect of some drug on blood pressure without knowing what it will do. If the R^2 is as high as 0.6, it's time to apply for a patent and contact the advertising agency.

The value of each R^2 can be determined almost instantly with a computer spreadsheet, which is a major convenience. It means that *c* can be changed and the effect observed faster than the eye can blink. It then becomes easy to try out different values and fine-tune to get the highest value for R^2.

R^2 can approach 1.0 but can never actually get there, unless there are only two data points so there is only one straight line. One reason is because most measurement are not exact unless they can be tallied by counting fingers. Stopwatches measure to an elapsed 0.1 second, but that still is not be good enough.

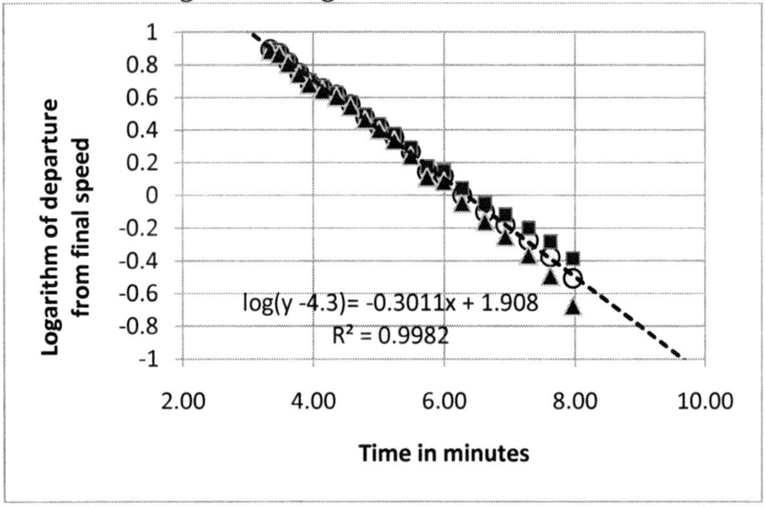

Fig. 4. Trial-and-error solution for the final speed of The Manhattan icebreaker crashing through a particular thickness of ice. The lines are for final speeds of 4.3, 4.2, and 4.4 knots. The4.2 knot speed gives the best straight line equation.

Getting back to reality

In Fig. 4 the picture is distorted by *logarithms* that can be brought back to the real world with *antilogarithms*. They are hidden away under a button labeled 10^x on some calculators. That changes the straight line in Fig. 4 into the curved line in Fig. 5, which is much less mysterious. It also shows a little bump after 8 minutes of constant plowing ahead into the ice. The bump means a slight change in conditions, most likely from encountering slightly thicker ice. It reset the clock for FORE. That is the topic of the next chapter. FORE is eclectic; the next chapter deals with pole vaulting.

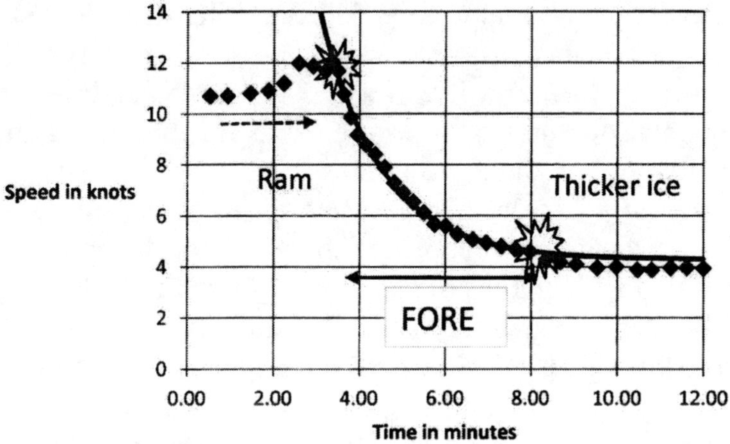

Fig. 5. Dots show actual measurements during a ram. The line goes through the dots and gives an ultimate speed of 4.3 knots but then the ship went into slightly thicker ice.

The ice party

Contrary to expectations, the ice party did not just go out and party on the ice. It was composed of six college engineering students and a professor who went along and did not plan to do anything. The students were there to drill holes in the ice and measure it strength and thickness. However, when the ship got stopped by the ice there seemed little point, so they took off their boots and parkas and headed for the food pantry.

When the ship backed and made a ram, the jolt woke the professor up at the very instant when he was contemplating Newton's Second Law, which is the one that he wrote to honor his mother, $f = ma$.

Newton's Second Law says that a rate of speeding up or slowing down, represented by a, can be multiplied by the mass m that is doing the speeding up or slowing down, to give the force f that is responsible for the change. In this case that should be the resistance from the ice, which was the exact answer that was needed. Since Russian scientists had tried it and it did not work, the students naturally tried it anyway, and it worked! The Russian icebreaker probably rode up on top of the ice, corrupting the data. The Manhattan did not do that.

Measuring speed.. The forward speed of the ship was measured with a sophisticated sonar system that got preoccupied with chunks of ice. The ship also had radar and GPS but they were not reliable. Dr. Andy Assur, a scientist with USA-CRREL (U.S. Army Cold Regions Research and Engineering Lab), had suggested a simple alternative, which was to stick a couple of boards out over the side and use a stopwatch to measure travel times for snow drifts to go the distance on the ice between the boards. The students wore out five stopwatches, which shows why it always is important to have a backup. Then the on-board computer failed, and the back-up was an antique "slide rule" that had been taken along as a joke. That demonstrates the importance of jokes. A 2½ month, multimillion dollar research effort came down to measurements using a couple of boards, stop watches, a portable tape recorder, pencil and paper, and a slide rule. It worked but it never would have made it to the moon.

A future for ice breaking?

The Manhattan is gone. A model is in the Smithsonian. The Northwest Passage is ice-free during summers and can be traversed with ordinary sailboats. There are even hotel accommodations scattered along the way.

Summer pack ice on Arctic seas now covers only about one-half of its former area, and even there it is only about half as thick. Melting already has opened the Northeast Passage that runs along the north coast of Siberia. Ice breakers still may make a comeback in order to extend the summer shipping season.

Secrets of the igloo. Years after the ice breaking data were collected, stored away and forgotten, they were pulled out to test the FORE equation. Then another spin-off was tried. It had to do with the shape of an igloo. It ended up with a picture that made the cover of Arctic magazine and a student edition of the Encyclopedia Britannica.

The classic architecture of cathedrals uses Gothic arches, which are circles with the middle parts cut out and the sides pushed together. The arches are simple to draw but when they get real they require flying buttresses to keep sides from bulging out and the tops from caving in, which was discovered the hard way. The igloo uses a *catenary* shape. It is the shape taken by a chain held at the ends. Flipped over, tension with no kinks becomes compression with no bulges, so walls are stable and require no further support. The catenary shape is embodied in the St. Louis Arch, the U.S. Capitol, and the Roman Pantheon, but the Eskimo did it first. It is the natural shape of arches in the Arches National Monument. Now about the only places to see flying buttresses are at football games, where they belong to the cheerleaders.

Fig. 6. Franklin's ships were held fast in sea ice in the Northwest Passage until they were abandoned three years later. Courtesy of Gjoa Haven Northwest Passage Historic Park.

Fig. 7. Engineering students keeping their cool on the bridge of the Manhattan, using a stopwatch to get ship velocities after radar and sonar gave up. Author photo.

Fig. 8. Track of the Manhattan. Students are making trail along the port (left) side of the channel. The professor had to go in the helicopter to take the picture.

Fig. 9. Igloo demo by Inuit at Pond Inlet, Numavut, Canada. The spiral makes a catenary like that of the St. Louis Arch.

Fig. 10. Prof. U-No-Who cutting a long shadow at -25° F. Photo courtesy of Nicholas Sutton.

Chapter 3. A Giant Leap for Mankind

Neanderthal did not have the sewing needle, so he had to walk bent over to keep his clothes from falling off.

The pole vault is a kind of assisted high jump, with the assistance coming from the pole. The pole vault is the highest that a person can jump under his or her own power in a single leap. When the pole suddenly was changed, it reset the clock for FORE. In the language of FORE it was a *major event.* It changed everything.

History according to Merlin. The pole vault was invented by accident during a jousting tournament when a jouster going at top speed on horseback allowed the point of his lance get too low so it jabbed into the ground. It flipped him up off of his saddle and high up into the air, then to come down in a crash of conflicted sheet metal. That also led to invention of the can opener. The experience was so shattering that some knights went off to hunt dragons in Egypt after receiving some misinformation concerning the origin of the Pyramids. When that didn't work out they went on the Crusades.

Tarzan re-invented the pole vault after his rope stretched while he was swinging from tree to tree over a bayou full of crocodiles, which also was the origin of the famous Tarzan yell. His rope swinging also kept slamming his body into tree trunks, because he couldn't quite get the hang of it.

Technique of the pole vault

A successful pole vault is accomplished by running as fast as a person can run while carrying a long pole held point first, jabbing the end of the pole into a hole in the ground, and holding on. As the pole swings upward, speed is converted into height. A stop-action photo at the end of this chapter shows that at the top of the vault the forward movement almost stops. On the way back down, height converts back into speed so it is important to have a soft landing. Early pole vaulters must have been unaware of the energy conversions because they went head-first and ended up wearing neck braces. Now pole vaulters are more sensible, go feet-first and land on foam rubber.

Competition. The objective of the competitive pole vault is to get the entire body over a horizontal bar without touching it, because if any part of the body touches the bar, it falls down. If no part of the body touches the bar, they raise the bar. That's life. Men wear a special adaptation of the sling shot to keep dangling from tangling.

The first recorded world record for the pole vault was in 1912, 13 ft 2¼ inches (4.02 m), using a solid wooden pole. Some vaulters tried to get a little extra height by shinnying up the pole, but it didn't work out. One little mistake and they were shish kabob.

Wood poles soon were replaced by bamboo that is strong, light and hollow, and heights kept improving. In the 1940s, Cornelius "Dutch" Warmerdam was the first person to go over 15 ft (4.57 m) using a bamboo pole. He did it in competition 43 times before anybody else did it once, so it looked like there must be a natural height limit that would make this a candidate for FORE. Warmerdam's final world record was 15 ft 7 ¾ inches (4.77 m). It was unequaled until after the introduction of different kinds of poles.

The importance of the pole. In 1957 Bob Gutowski broke Warmerdam's record using an aluminum pole, and in 1960

Don Bragg broke that record using a steel pole. Then fiberglass poles were introduced and records really started tumbling. By then Warmerdam was retired from vaulting, but suggested that he could have gone 2 1/2 feet higher if he had had a fiberglass pole. Was he right, or was it only because he lived in California? Might he still hold the world record? FORE can look into it.

The big bend

Fiberglass already was being used in fishing poles because fishing is cutting-edge. As shown in a photo at the end of this chapter, a fiberglass pole bends like a spring. It stores energy like a spring, energy that is released and can throw the body higher when the spring straightens out. It works like a bow-and-arrow. The bend of the bow is what makes the arrow go.

Neanderthal man did not have the bow-and-arrow. He used a spear. He also did not have the sewing needle so he had to walk bent over keep his clothes from falling off. He used one hand to hold his spear and the other hand to grab his clothes so he could not have used a bow-and-arrow even if he had one. Then he drowned in Great Flood because he couldn't swim with one hand. Remember, you saw it here.

Fig. 11 shows how record heights in the pole vault were leveling off and then shot upward after changing the pole. Fig. 12 shows the same data analyzed with FORE. Ultimate heights were varied and as discussed in the preceding chapter, the values were selected that give the best straight lines. In Fig. 13 the logarithms were brought back into the real world.

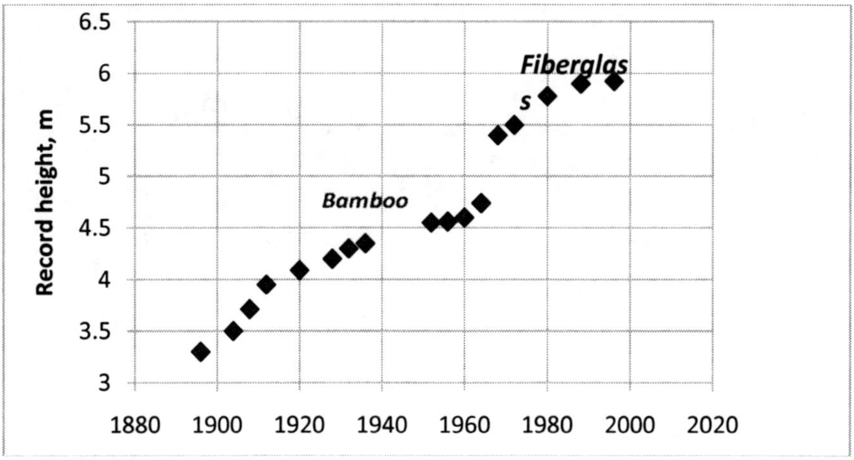

Fig. 11. Olympic record heights in the pole vault. Something changed after fiberglass was substituted for bamboo in the pole. In the language of FORE it was a major event.

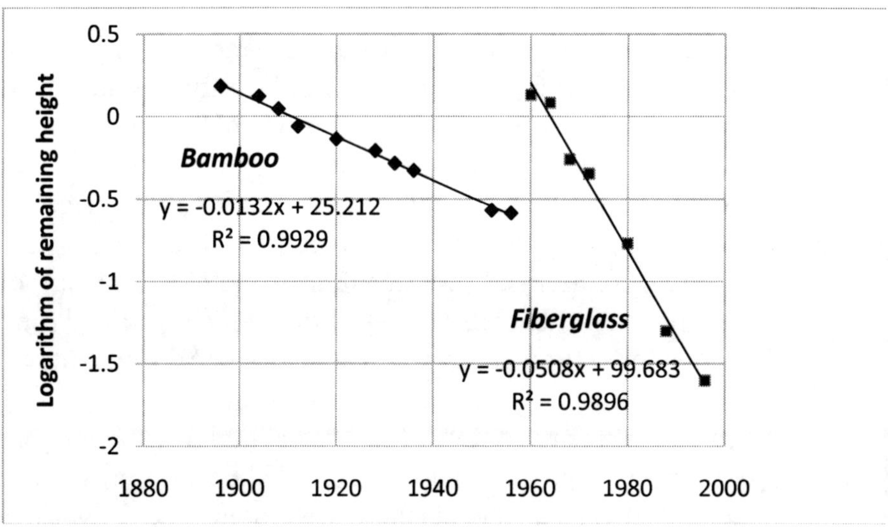

Fig. 12. Two straight lines and two formulas determined by FORE for ultimate heights with the two kinds of poles. There really is a difference.

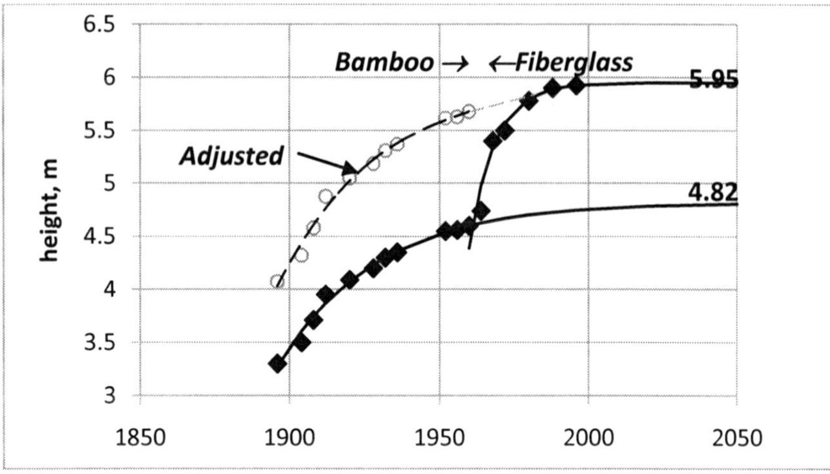

Fig. 13. Lines drawn with FORE go through all of the data points, which is consistent with the high R^2 values. The "adjusted" line shows fiberglass equivalents based on the ratio of ultimate values.

FORE gives maximum heights of 4.82 and 5.95 m (15 ft 9.8 inches and 19 ft 6.25 inches) for bamboo and for fiberglass poles. That represents an increase of 23 percent. It probably is one of the most drastic changes in all sports records, a shot for the stars. There was absolutely no way that bamboo poles could accomplish that. Fiberglass turned the page, and the page got turned not as a result of athleticism but because of better engineering. It still took a while for vaulters to take full advantage of the advantage.

Old vaulting records may be ancient history but they can get a reprieve by being translated into fiberglass equivalents, which can be accomplished by multiplying 5.95/4.82 = 1.23. The dashed line in Fig. 13 shows the corrected bamboo records, and the line precisely merges with actual records so it looks like something must be right.

World versus Olympic records

Figures 11-13 are for Olympic instead of world records because Warmerdam's achievements were so outstanding that they put a spike in the world record curve. That is not unusual, and designates what may be called a "superstar," one who is ahead of his time. He did not compete in the Olympic Games because of WWII. Fig. 14 shows FORE analyses for world records, and although the R^2 values are lower, the result is the same, a height advantage of 23 percent from the use of a fiberglass pole.

Might Warmerdam have gone over 18 ft? Adding 23 percent to Warmerdam's final world record of 15 ft 7 3/4 inches gives 19 ft 4 1/2 inches. Not only would he have gone 18 ft, he would have had another foot-and-a-half to spare!

Might Warmerdam still hold the world record? After the adjustment his record would not be broken for 42 years, which in itself is a record. It was broken in 1984 by Thierry Vigneron of France, who then held the record for 10 years. It then was broken by Sergei Bubka of Ukraine, who certainly is one of the best pole vaulters of all time. He has held the world record for over 20 years and counting. As a world record approaches a FORE ultimate, there is less room left for improvement.

So what is it with fiberglass? Fiberglass poles consist of very fine glass fibers laid parallel and embedded in plastic. The fibers derive their strength from *surface tension,* which is what makes soap bubbles round and lets water bugs walk on water. Carbon fibers now are used in fishing poles, but are regarded as a bit too whippy by many pole vaulters.

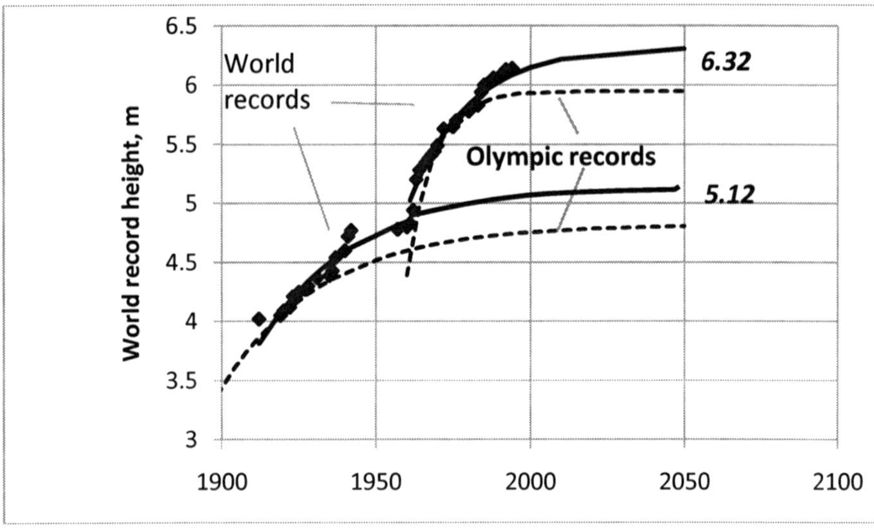

Fig. 14. World records are less consistent than Olympic records but give the same 23 percent increase in height from the use of fiberglass poles.

Table 2. Men's world records in the pole vault. The dashed line shows Warmerdam's reign after adjustment is made for a fiberglass pole.

Date	Record Holder	Measured	Fiberglass equiv.
1912	M. Wright, USA	4.02 m	4.94 m
1920	F. Foss, USA	4.09	5.09
1924	R. Spearow, USA	4.22	5.19
1927	S. Carr, USA	4.27	5.25
1928	L. Barnes, USA	4.30	5.29
1931	C. Hoff, Norway	4.32	5.31
1931	W. Garber, USA	4.37	5.38
1935	K. Brown, USA	4.39	5.40
1936	G. Varoff, USA	4.43	5.45
1937	B. Sefton, USA	4.54	5.58

Table 2 (Cont.)

Date	Record holder	Measured	Fiberglass equiv.
1937	E. Meadows, USA	4.54	*5.58*
1942	C. Warmerdam, USA	4.77	*5.87*
	End of bamboo pole competition		
1957	B. Gutowski, USA	4.78	--
1960	D. Bragg, USA	4.83	--
	Beginning of fiberglass pole competition		
1961	G. Davies, USA		4.83
1962	J. Uelses, USA		4.89
1962	D. Tork, USA		4.93
1962	P. Nikila, Finland		4.94
1963	B. Sternberg, USA		5.08
1964	F. Hansen, USA		5.28
1967	R. Wilson, USA		5.38
1969	J. Pennel, USA		5.44
1970	W. Nordwig, E. Germany		5.48
1970	C Papanikolaou, Greece		5.49
1972	K. Isaksson, Sweden		5.59
1972	B. Seagren, USA		5.63
1976	E. Bell, USA		5.67
1976	D. Roberts, USA		5.65
1980	P. Houvion, France		5.77
1980	W. Kozakiewicz, Poland		5.78
1981	V. Polyakov, USSR		5.81
1983	P. Quinon, France		5.82
1984	T. Vigneron, France		5.91
1994	S. Bubka, Ukraine		6.14
	Ultimate from FORE		6.32

Here come the girls!

Women were late bloomers in the pole vault, most likely because their bloomers kept catching on the bar. Now things are more out in the open. World record keeping in the women's pole vault started in the 1970s, but it was not until 1990 that it became an Olympic event.

Women's records have gone up in steps that show improving performances of individual athletes, Fig. 15. The current world record holder is Yelena Isinbayeva of Russia, who has broken her own world record over a dozen times. Her current best, 5.06 m (16 ft 7¼ in.), is almost a foot higher than Warmerdam's best with a bamboo pole. However, it is over a meter (4 ft) short of Bubka's record with fiberglass, so it appears that there really is a "gender factor." So what is it?

"Normalizing." Men usually are taller than women, but there is a way to take height out of the equation. Bubka's highest pole vault is a little over 3.3 times his height, and Yelena's best so far is 2.9 times her height. That means that height is not the only difference. This method of data processing is called *normalizing,* which means to make it incomprehensible.

Since height must not be the only difference, what is the secret from being male? The FORE ultimate record for women is 5.37 meters, Fig. 15, and for men it is 6.32 meters, Fig. 14. That indicates an 18 percent advantage from growing up male. Removing the height factor still leaves 14 percent advantage. It still shows that men really are different from women. The FORE analysis for women's records is a bit shaky because the time span is short, and the R^2 value is nothing to brag about, 0.98, but that is not enough to affect the difference between male and female.

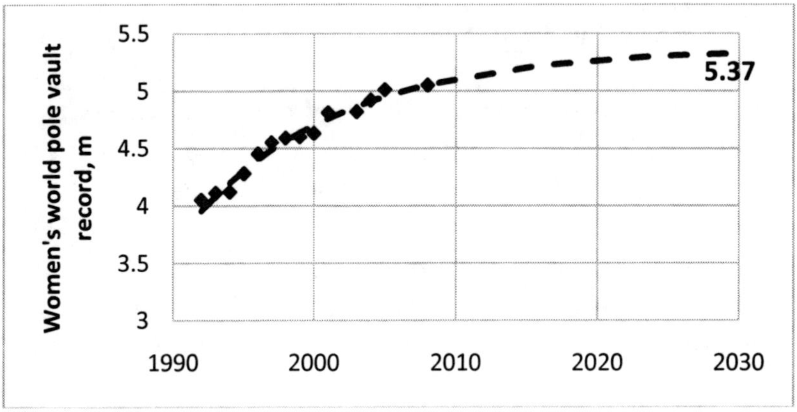

Fig. 15. Women's world records in the pole vault with the fiberglass pole. The FORE projected ultimate height based on a relatively short time span is 5.37 m (17 ft 7½ inches) with $R^2 = 0.98$.

Table 3. Peak performances of women's world record holders in the pole vault, fiberglass poles.

Date	Record holder	Height
1995	Sun Caiyun, China	4.08m
1995	Andrea Müller Germany	4.18
1995	Daniela Bártová, Czech Republic	4.22
1999	Emma George, Australia	4.60
2001	Svetlana Feofanova, Russia	4.64
2001	Stacy Dragila, United States	4.81
2009	Yelena Isinbayeva, Russia	5.06
	Ultimate from FORE	5.37

Extra credit for math majors:

Men tend to be heavier so what about weight; does it make a difference? It can make sense that lighter can mean going higher, but does it make good sense?

Ans. There are two general classes of energy that affect the pole vault: *Kinetic energy* is the energy of speed, and *potential energy* is the energy of height. The pole vault mainly converts the two kinds of energy.

Kinetic energy is: $E_k = ½\ m \times v^2$

so it is heavy into velocity. Potential energy is:

$E_p = m \times g \times h$

The *m* in both equations mass, which relates to weight. Dividing both sides by *m* takes it out:

$½\ \cancel{m} \times v^2 = \cancel{m} \times g \times h$

which gives $v^2/2 = gh$

Dividing by *g*, which represents gravity, gives :

$h = v^2/2g$

Speed is the thing, if g, or gravity, is constant.

Fig. 16. Photo made in the late 1800s by French physiologist Etienne-Jules Marey. His strobe light consisted of a rotating slotted disk in front of his light source. A pole vault converts speed into height that then converts back into speed: The closer the spacing in the photo, the slower the speed.

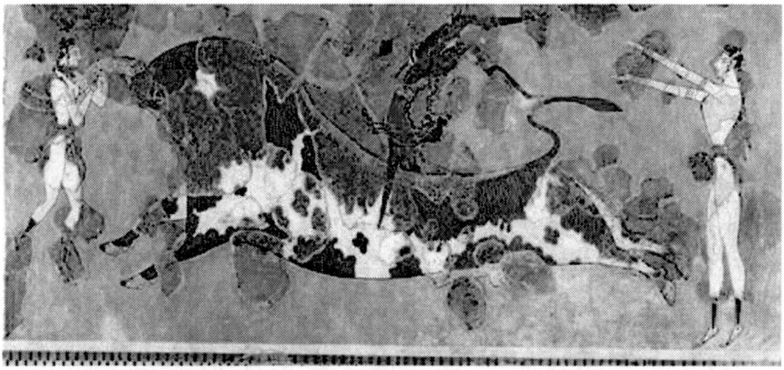

Fig. 17. Bull leaping: A 3500-year old fresco in the Palace of Knossos on the island of Crete shows girls grabbing a bull by the horns and vaulting over its back. When boys got involved they decided to play it safe and kill the bull.

Fig.18. The fabulous Yelena Isinbayeva, holder of the female world record in the pole. Thank you, Russia. http://www.yelena-isinbaeva.com

Chapter 4. Berlin 1936

22,000 pigeons fluttered upward over the stadium while numerous guns fired salutes, which the pigeons apparently took personally because they fired back.

On August 1, 1936, Adolf Hitler proudly acknowledged a rolling wave of *"Sieg Heils"* as he entered the new Olympic Stadium in Berlin. Now the world would bear witness to the New Germany and superiority of the "Aryan" race. This also demonstrated his flexibility because "Aryan" does not mean blonde and blue-eyed; it is from Sanskrit and designates a prehistoric tribe in Iran and Pakistan. Sometimes is politically expedient to be flexible before being rigid.

Richard Strauss of *Rosenkavalier* fame conducted music of Wagner, Hitler's favorite composer, as the mighty zeppelin the Hindenburg cruised over with huge swastikas splashed across its tail fins. Less than a year later it famously burst into flames at Lakehurst, New Jersey and to save face Hitler invaded Poland and started WWII. It was one huge mistake.

Hitler's rise to power had been aided by a group of beer-guzzling thugs called the "Brownshirts." He later had them shot for their decadence because they partied wearing lipstick and garter belts. Hitler then broadened his criteria to include Jews, Communists, intellectuals, Gypsies, religious figures, and all others who were not above suspicion. In ten years millions would be dead and Europe would be in ruins, much of it with a red flag flying over it.

Enter J. C. Owens.

As soon as an American athlete started removing his warm-up suit the crowd began clapping and chanting his name, "Jess-ay Oh-vens." Meanwhile Hitler was overheard muttering that America should be ashamed for allowing itself to be represented by blacks and Jews. Jesse Owens equaled the world record for the 100-yard dash while still in high school. At Ohio State University he was the "Buckeye Bullet." Ironically did not even have an athletic scholarship, but worked part-time jobs to pay for his education.

Owens was 22 years old when he accomplished what many experts regard as a defining moment in sports. In 1935 at the Big Ten Championships in Ann Arbor, Michigan, and after a painful fall where he hurt his back, he broke three world track records and equaled a fourth, all in less than an hour! The new records were in the 220-yard sprint, the 220 yard low hurdles, and the broad jump, now called the long jump. The same day he tied the world record in the 100-yard dash. Arguably he also would have broken two more world records, in the 200-meter sprint and hurdles, had those shorter distances been timed. A few months later he set a new world record in the 100-yard dash.

Fig. 19. Jesse Owens off to a good start, leaning forward and arms pumping. He was running on a cinder track.

In the Berlin Olympics, Owens won gold at 100 meters, running on the inside lane that was the muddiest part of the track. The silver medal went to another outstanding black athlete from the U.S., Ralph Metcalf, who later did his successful running for Congress. Bronze went to yet another American athlete, Frank Wycoff, which made it a clean sweep for the U.S. The U.S. team was a heavy favorite to win gold in the 4×100 meter relay, where four runners each run 100 meters, and the U.S. team easily won but there was criticism because at the last minute Owens and Metcalf had been substituted for two other runners who also almost certainly would have won gold. They were Sam Stoller and Marty Glickman. They were not black; they were Jewish. Owens had refused to run and was told that he had no choice.

Hitler decided to avoid any embarrassing photo-ops by not shaking hands with *any* of the winners, although he did manage to wave a hand in Owens' direction before an approving crowd. The leading politician on the other side of the pond also did not personally congratulate Owens because he was worried about losing votes in the "Solid South." A few decades later the Solid South did solidly switch over as a result of the civil rights movement.

Jesse Owens died in 1980 from lung cancer. He was 66. He smoked. At that time actors posing as doctors were hawking cigarettes, and cigarette companies were supporting higher education by handing them out free to students on college campuses. Although cigarettes kill more people than wars and pestilences, that is much to the regret of the tobacco companies, because dead people don't buy cigarettes. That is called "corporate empathy." Their only way out was to promote overseas sales.

Foiling the Nazis

The German Olympic team included a tall, striking blond who was an expert with the sword. She also fit the image of the ideal "Aryan" superwomen except that she was half Jewish.

Helene Mayer was studying and teaching at Mills College in California when she got the call to participate for Germany in the Olympic Games. In the 1928 Olympics at the age of 17 she had captured gold in fencing, so it in the 1936 games it was convenient for Germany to temporarily overlook her troubling ancestry. She then showed a rapier-like mastery of the situation by demanding that her German citizenship be restored even though she had no plans to use it. It would have been embarrassing to the German government to reveal why her citizenship had been revoked in the first place. Helene won silver at the Berlin Games, and gold went to a Hungarian named Ilona Elek. She also was half-Jewish.

Mayer was criticized for giving the Nazi salute while standing on the podium to receive her medal, but her

Fig. 20. Helene Mayer's famous salute, understandable because her family was Jewish and still living in Germany. She later became a U.S. citizen.

family still lived in Germany and she knew the score. She returned to the U.S., changed the spelling of her name to Meyer, and became a U.S. citizen and a popular teacher.

"The world's fastest human"

One of the more colorful runners ever to grace the Olympics was Charlie Paddock, who liked to call himself "the world's fastest human." At the 1924 Olympics he took gold in the 100-meter, and a year later trimmed 0.2 second off of the world record to take it down to 10.4 seconds.

Charlie liked to add a little drama at the finish by leaping forward with his arms outstretched to celebrate his triumph. Pushing the arms forward meant pushing the body back but he won anyway. His grandstanding was a bit much for the Olympic Committee at a time when being an amateur really meant being an amateur, and his application was challenged for the 1928 games. He was defended by the committee chairman, Gen. Douglas MacArthur, who also did not mind doing a little grandstanding. Charlie wrote a scathing article for *Colliers* magazine and was depicted as the brash American in the 1981 movie, "Chariots of Fire." During WWII as a U.S. Marine he was killed in an airplane crash in Alaska.

Fig. 21. Charlie Paddock at the finish. Some people regarded this as a little extreme.

How it all got started

Times were not measured at the first Olympic Games, which were held in 776 BC in Olympia, Greece. That probably was because of limitations of the sundial. Runners used a standing start, and since the weather was hot and the men were cool they ran in the nude. Married women were not allowed to watch. Unmarried girls were allowed but were discouraged because of the giggling.

Competitors ran the length of a field called a *stadion*, which equaled about 180 m, or a bit short of 200 yards. Athletes slicked up with olive oil. When it got dirty they used special curved sticks to change the oil. Even today body builders slick up their bodies with oil so when they try to make love they slide off the bed.

Fig. 22. Practice session for the Olympic torch lighting ceremony in Olympia, Greece. Author photo.

The double stadion. In another race, runners ran to the end of the field and then ran back, making it possible to run hard without getting anywhere. The longer distance was supposed to represent the range of Persian arrows but that seems a little shameless. In 490 BC at a famous battle at Marathon, Greece, 10,000 Greek soldiers stood their ground against the Persians, and a soldier with a Greek name (we don't do Greek names) ran all of the way from Marathon to Athens to give the good news. He then dropped dead from exhaustion. That was the origin of the marathon but now the runners only feel like they could drop dead.

After more than a millennium the Games were cancelled in 393 by a Roman emperor when he got the Faith and concluded that they were pagan rituals. He also disbanded the famous Roman Vestal Virgins, but there also might have been a shortage of virgins. For the next 1500 years there were no Olympic Games, not until 1896 when they were reinstated in Athens. The Olympic Flame came back in the 1920s. It now is lit with natural sunlight focused with an unnatural lens, at the site of the original Games at Olympia.

How the standing start ended

In the 1896 Games a student from Boston University named Tom Burke crouched down so everybody probably thought he must be tying his shoelaces, but he knew what he was doing. At the instant that the race got started, his legs took turns pushing like pistons to give him a head start. He won the 100-meter and was awarded the first gold medal given in a modern Olympic Games. He then won gold in the 400-meter, and that sealed the deal; the standing start was reserved for long races where too much energy would be used to stand up.

Paved vs. cinder tracks

Cinders were soft and cheap compared to concrete that was hard and expensive, so cinders were widely used for running tracks. The cinders usually were available free from coal-burning municipal power plants. Runners would scrape out holes for their toes to get a good push-off at the start, and track shoes had spikes that dug in and prevented slipping. However, the spikes also had a disadvantage because when they dug in and scuffed things up they used energy. Starting in the 1960s running tracks started to be paved, and now special materials are used to make them resilient and give a little extra bounce. Starting blocks are used for the initial push-off. Today's running tracks often have a surface layer of polyurethane or asphalt with chips of rubber from recycled tires, and a little extra bounce can be achieved with an overlay of rubber chips held together with asphalt or latex.

Might those changes qualify as a major event like changing the pole in pole vaulting? If so, how might that have affected world records?

Fig. 23 shows declining world record times for 100 meters. The line is empirical, which means that there was no guiding principle except to go close to or through the data points, and the R^2 value of 0.95 is pretty good. So how might the empirical equation project into the future?

A continuation of the empirical curve indicates that by 2020 the record time will be down to zero, highly unlikely, and after that it will be negative, which would mean that a runner would have to finish before he started. About the only place that can happen is in science fiction, with the emphasis on the fiction. The lesson is that an empirical equation only applies to the range represented by the data; it can't be reliably extended except on the talk shows.

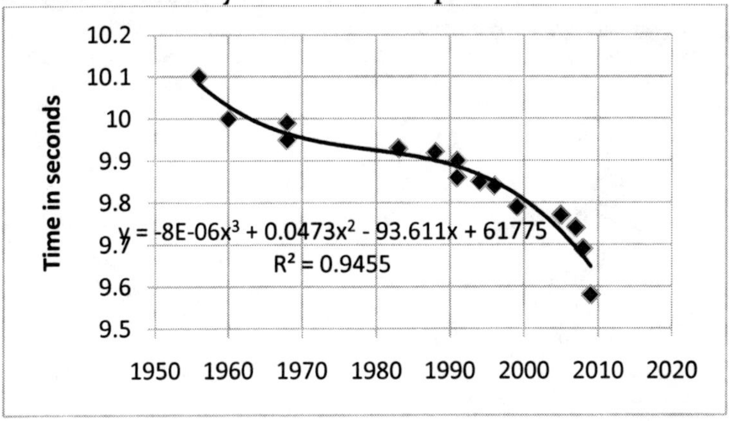

Fig. 23. According to this graph and equation eventually runners will have to go backward in order to catch up. These things can happen with empirical equations.

The starting gun

A starting gun can only shoot blanks because the intent is to make a sharp noise without killing anybody. Scientific tests show the superiority of a starting gun over all other methods, including "Ready, Set, Go!" The effectiveness also was demonstrated when Billy the Kid entered a saloon and was told he was too young to have a beer.

Sound travel takes time, so in order to be fair a microphone is held close to the gun and the message transmitted with loud speakers.

Perils of stopwatch timing

Stopwatch timing depends on human reaction times and rheumatism of the thumb. Three official timers usually were used, sometimes with three backup timers in case a timer blinked, sneezed, or was not paying attention. If the three official times did not agree, as a conservative measure the slowest time often went into the record. Statistically it would be better to use an average, but a real world record can't be an average.

Fig. 24. Obsolete timing device available at garage sales.

Stopwatches measure time in tenths of a second, and Owens' record time of 10.2 seconds in the 100- meter was equaled ten times in the next 20 years, but was not broken. Then in 1956, four runners competing in various events, Willie Williams, Ira Murchison, Lemon King and Ray Norton, all took the record from 10.2 to 10.1 seconds. Between 1960 and 1969 ten runners posted 10.0 seconds flat, partly because they the advantage of a new track, and it became popular to refer to a "10-second barrier."

How to get the signal. With stopwatch timing a substantial fraction of a second is required for a visual stimulus to go from eyeballs to the brain, where it is processed to compose a message that has to go all of the way through an arm and activate a thumb. In theory the time should be the same at the start and at the end of a race, but there still can be error at both ends of the trail.

Electronic timing and photo finishes have improved measurement precision by a factor of 10, to the nearest 0.01

second. That is about one-third of the time it takes for the blink of an eye. Now with digital photography there is no waiting for film to be developed.

Working with major events

Fig. 25 shows the same data as Fig. 23 but with record times allocated according to three technological advances: First, the world records for 100-meters set on cinder tracks; second, those set on paved tracks; and third, those set on paved tracks using electronic timing.

Influence of paved track. The FORE best-fit straight lines give ultimate world record times of 10.03 and 9.87 seconds for running on cinders compared with on paved tracks. Although that is only a difference of 1.6 percent, that still is enough to defeat the "10-second barrier" for cinder tracks.

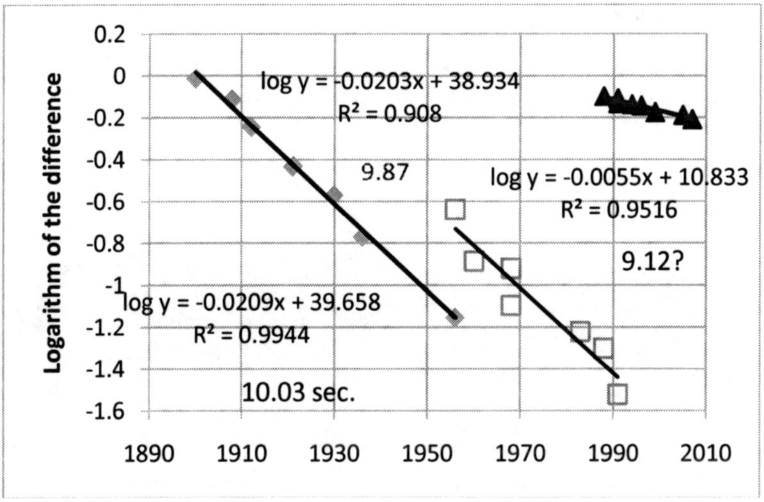

Fig. 25. Introduction of paved track was a major event, as FORE ultimate times went from 10.03 to 9.87 seconds (the two lines at the left). A third interval started in 1989 by electronic timing may not be a reliable projection.

At the 1968 Olympic Games in Mexico City, the track was covered with a resilient polyurethane plastic with the trade name "Tartan." Spikes were allowed, but could be no longer than ¼ inch (6 mm) and had to be dull so they would not dig in. The "10-second barrier" no longer was a barrier. Jim Hines broke through with a stopwatch time of 9.9 seconds. A few months later he posted an electronically measured time of 9.95 seconds.

The conversion from logarithms back into real time is shown in Fig. 26. Paved track may only have made a small difference, but it helped to slip past the 10-second barrier and break some world records. Paving the track was a *major event*.

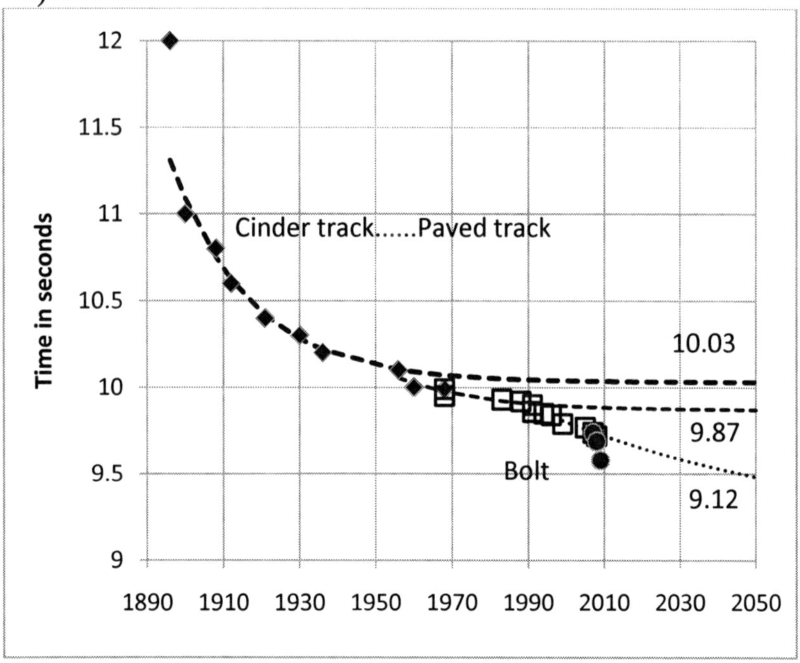

Fig. 26. FORE ultimate records for cinder and for paved tracks, and a questionable projection for paved track and electronic timing as the time span is too brief.

Might Jesse Owens still hold the world record at 100 m? Jesse Owens' best time of 10.2 seconds translates into 10.04 seconds on paved track, not enough to retain a world record. However, he would have held the record for 20 years, which can be taken as a measure of a "superstar." Jim Hines held the world record for 15 years.

Hot timing

Electronic timing has greater precision than stopwatch timing and arguably should be more accurate. The FORE ultimate with stopwatch timing was 9.84 seconds. Then in 1979 with electronic timing Maurice Greene took the record down to 9.79 seconds. In 2005, Asafa Powell shaved off another 0.02 second and two years later took off another 0.03 second. Then Usain Bolt took record down to 9.72 seconds, well below the projected ultimate from stopwatch timing.

The big surprise came in 2009 when Bolt posted a phenomenal 9.58 seconds for 100 meters. A FORE projection of 9.12 seconds in Fig. 22 is questionable; the line it is straight but it covers only a brief time span and includes a superstar, Usain Bolt.

Is there a height factor?

Usain Bolt is tall, 6 ft 5 in. (196 cm). His fellow Jamaican, Asafa Powell, is 6 ft 3 in. (190 cm). Longer legs take fewer steps, but a larger body mass takes longer to get started. Usain Bolt is not known for his fast starts.

Model theory. Most models are not particularly strong on theory except for applying makeup. A scale model does not mean a model who steps on the scales, but has to do with reduction in size. Fig. 24 is a graph of record times in the 100m dash plotted against and heights.

Table 4. Progression of men's 100m world track records

Date	Record holder	Cinder	Paved	Years
1912	D. Lippincott, USA	10.6 sec	10.4	13
1921	C. Paddock, USA	10.4	10.2	9
1930	P Williams, Canada	10.3	10.1	6
1936	J Owens, USA	10.2	10.0	20
1956	W. Williams, USA	10.1	9.9	3
1959	A Hary, W. Germ	10.0	9.8	11
FORE	*ultimate for cinder*	*track*	*10.03*	
1968	J Hines, USA		9.9	15
	Electronic timing			
1968	J Hines, USA		9.95	—
1983	C Smith, USA		9.93	5
1988	C Lewis, USA		9.92	3
1991	L Burrell, USA		9.90	0
FORE	*ultimate for paved*	*track*	*9.87*	
1991	C Lewis, USA		9.86	3
1994	L Burrell, USA		9.85	2
1996	D Bailey, Canada		9.84	3
1999	M Greene, USA		9.79	6
2005	A Powell, Jamaica		9.77	2
2007	A Powell, Jamaica		9.74	—
2007	U Bolt, Jamaica		9.72	6+
2009	U Bolt, Jamaica		9.69	—
2009	U Bolt, Jamaica		9.58	—
FORE	*ultimate, paved track,*	*electronic*	*timing*	*9.12?*

Maurice Greene's ambition since childhood was to be the fastest person in the world, and at 5 ft 9 inches (170 cm) he did not let a short stature get in his way. He broke 10 seconds 53 times during his career, and his fastest time of 9.79 seconds remained the world record for 6 years. What if he had been taller? It is pure speculation, but the dashed line suggests that if he were as tall as a Bolt, he might even have run like a Bolt.

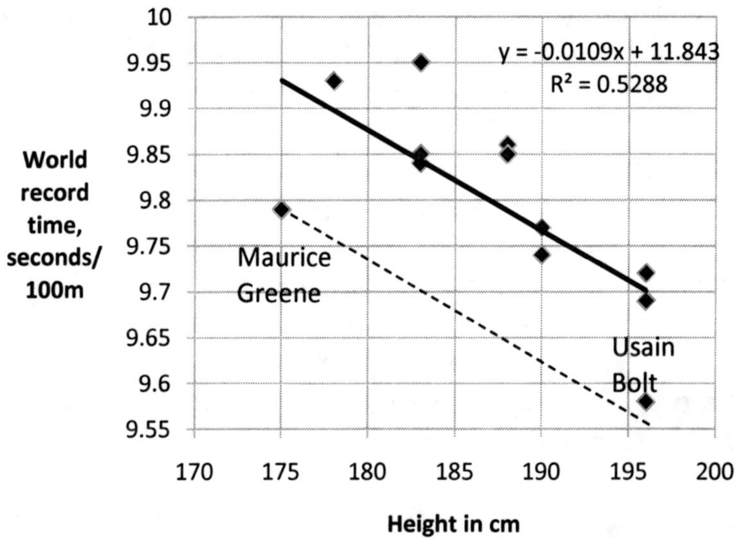

Fig. 27. A graph that has nothing to do with FORE but shows a fairly consistent trend, that record times partly depend on heights of the runners, probably because tall means fewer steps.

The 100-yard dash

For years the 100-yard dash was a sacred fixture in U.S. track meets because it is the length of the American football field. Since 100 yards is only 91.44 meters, record times are shorter.

In 1973 Ivory Crockett set the still-extant world record of 9.0 seconds for 100 yards. That is so phenomenal that it is too bad that the race did not continue for another 28 feet (8.56m), because if he had been able to keep up the same pace he might have finished 100 m in 9.85 seconds, which also would have been a world record. A year earlier he actually did run 100 meters in even less time, 9.6 seconds but the record was set aside as "wind-aided." There also may have been some reluctance to believe such an unbelievable world record that would not have been broken until the best of Bolt, and Crockett was even shorter than Greene, at 5 ft 7 in. (170 cm). A park is named after him in his home town of Webster Groves, Missouri.

Fast and female

Women's world records at 100 meters are shown in Fig. 28 and are a fair fit for FORE. According to this, Helen Stephens and Florence Griffith Joyner or "Flo-Jo" were ahead of their times, Flo-Jo spectacularly so. She was tall at 5 ft 7 in. (170 cm), leggy and glamorous, and liked to dress the part. In 1988 when she took 0.27 second off of the world record it drew some sparks even though the cut was less than that made by Stephens. One claim was that the wind gauge was not working but winds do vary. In 1988 Flo-Jo also set and still holds a world record at 200 meters, wind gauge working.

The FORE analysis is in Fig. 29, and Flo-Jo's record is not included. She vigorously denied accusations of drug use, and credited her rapid improvement to body-building exercises regimented by her husband, coach, and fellow Olympic champion, Al Joyner. So was her performance within the realm of possibility?

The answer is yes. From the other performances, the FORE projection for an ultimate time is 10.47 seconds, 0.02 second less than Flo-Jo's best time but it was cutting it close. It can perhaps be put in better perspective by Charlie Paddock's record of 10.4 seconds that was set on a cinder track in 1921. Record times are converted to the equivalent speeds in Fig. 30, and men really can run faster. The ratio of ultimate times for men and for women is 9.12

÷ 10.47 = 0.87, with men 13 percent faster. The ratio of actual world records for 100 m, those of Joyner and of Bolts, is 9.58 ÷ 10.49 = 0.90, with men 10 percent faster.

Flo-Jo tragically died in her sleep at the age of 39, the victim of a seizure.

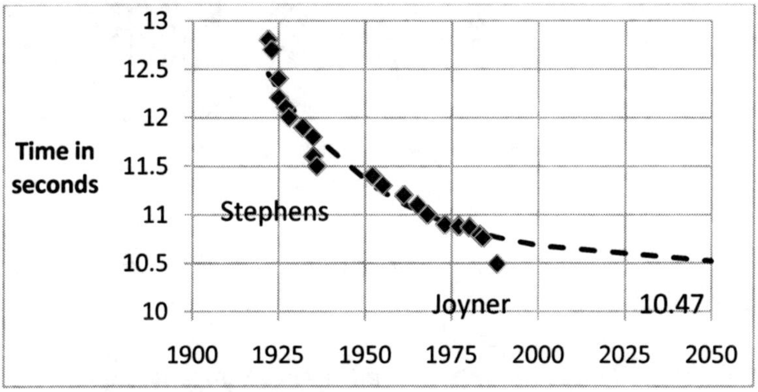

Fig. 28. Women's world records in the 100 m. The curve was drawn with FORE and indicates that Flo-Jo was years ahead of her time.

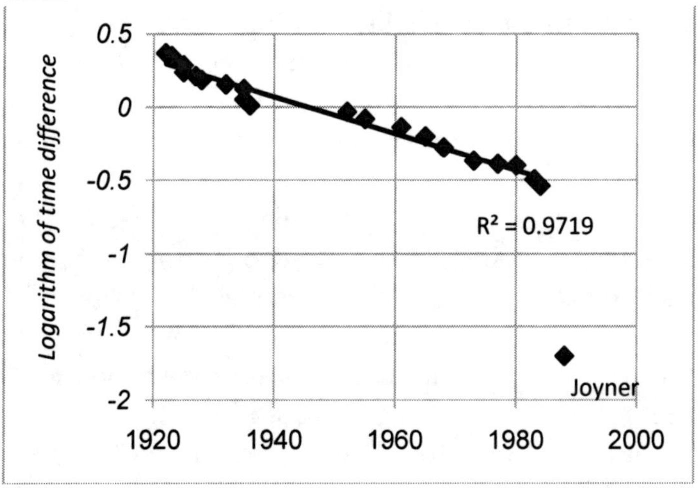

Fig. 29. FORE gives a high R^2 by excluding Joyner's record, and gives a projected ultimate value for women of 10.47 seconds.

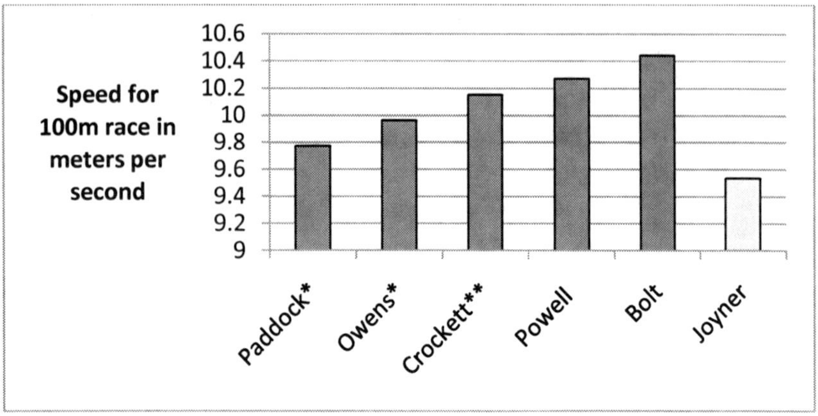

Fig. 30. Boys really can run faster. It may be inherited from caveman days when they had to run faster to in order to catch girls.
**Corrected for cinder track.*
*** Corrected and converted from the record for 100 yards.*

Which race is faster?

Record times for 100 meters can be converted into average running speeds by dividing them into 100. That turns seconds per 100 meters into meters per second. The same procedure can be used for other lengths of races. More fatigue and shorter races put more emphasis on the time to get out of the starting blocks.

Florence Griffith-Joyner's world record speeds in the 100 and 200 meter are 9.53 m/s and 9.37 m/s respectively, indicating that in her case fatigue played a larger role. In both cases 100 meters is the fastest race. Shorter races often are run indoors and put more emphasis on the start-up time, and average speeds are slower.

Usain Bolt holds the men's record speed for the 100-meter at 10.44 meters per second, which corresponds to 24.3 miles per hour. He also holds the world record speed at 200 meters, where his average speed is only slightly slower, 10.42 meters per second.

Table 5. Progression of women's 100 meter world track records.

Date	Record holder	Time	Years held
1922	M Mejzlikova, Czec	13.6 sec	—
1922	M Lines, UK	12.8	3
1925	L Schmidt, Germany	12.4	1
1926	L Junker, Germany	12.2	2
1928	B Robinson, USA	12.0	4
1932	T Schuurman, Neth	11.9	—
1932	S Walasiewicz, Pol	11.7	3
1935	H Stephens, USA	11.6	17
1936	H Stephens, USA	11.5	—
1952	M Jackson, Australia	11.4	4
1955	S Strickland, Aust	11.3	5
1961	W Rudolph, USA	11.2	4
1965	I Kirzenstein, Pol	11.1	3
1968	W Tyus, USA	11.0	5
1973	R Meifsner, Germany	10.9	2
	Electronic Timing		
1977	M Oelsner, Germany	10.88	6
1983	M Göhr, Germany	10.81	—
1983	E Ashford, USA	10.79	4
1984	E Ashford, USA	10.75	—
1988	F Griffith-Joyner, USA	10.49	25+
	FORE ultimate	10.47	

Designer feet

Feet put spring in the step and keep things moving, but there is a cost. A foot on the ground is stationary, which means that the other foot has to go forward twice as fast in order to make it up and delay. Meanwhile the rest of the body keeps moving at a steady pace. It is the stop-and-go of the running apparatus that uses the energy, and big feet or heavy shoes do not help. Distance runners in particular have thin legs, plus strong hearts and breathing apparatus. Some even run barefoot. Some of the thinnest feet belong to birds that also happen to be good runners.

A wheel is more efficient because it can roll without stopping.

Fig. 31. Ron and Flo-Jo, two people who really knew how to run but in different venues. Courtesy of the Ronald Reagan Library, Sun Valley, California.

Chapter 5. Bats and Balls

The umpire stands behind the catcher and calls balls and strikes by pointing a finger up in the air. That is to show where he gets his information.

America is red, white, blue and baseball. When Babe Ruth hit record-breaking home runs, when Jackie Robinson turned jeers into cheers, that was America. Whether played in Cuba, China, Japan, Lapland, Belarus or Kyrgyzstan, baseball is a flavor of America.

One of the most famous baseball players of all time was George Herman Ruth, better known by his nickname "Babe." He began his career as a pitcher, and he was good at it. He was a left-hander, and left-handed pitchers can throw a bit of fear into right-handed hitters because the ball can look like it is coming right at them.

Fig. 32. Archival photo of Babe Ruth in his favorite stance.

Pitching a baseball is one of the most strenuous activities in sports, so a pitcher normally only pitches about every fourth game. Pitchers concentrate on their pitching so they usually are not good batters. When it was Babe's turn at bat he often was substituted by a "pinch hitter," but in 1918 as a pitcher he hit 11 home runs. The following year he hit 29, and his pitching days were over because he was such a good hitter. The Boston Red Sox then sold his contract to the New York Yankees, and Red Sox fans still are on a crying jag because Babe then set a new home run hitting record, 54 home runs in a single season. The following year he advanced the record to 59, and six years later, in 1927, he hit 60, a record that remained untouched for decades.

Making sense out of baseball

Baseball is not like other games because time usually is of no consequence. A game has nine innings with each team having its turns at bat, and after nine innings if the score is tied they keep adding more innings until there is a winner. In 1981 a game between the Pawtucket Red Sox and the Rochester Red Wings went 33 innings and lasted all night but it still was not over. The game was resumed a couple of days later, by which time losing did not look like such a bad option. The Red Wings lost.

The pitcher. A key figure in baseball is the pitcher, who throws the baseball in such a way that the batter has difficulty hitting it. Every time that a batter swings at a ball and misses, it a strike. Three strikes and the batter is out, which means that he has to go sit in the "dugout" and contemplate his shoe laces.

In order to create an element of fairness, the pitcher is required to throw the ball so it goes through a "strike zone" in front of the batter, even though behind would be more effective. If the ball is not hit it is caught by the catcher, who is behind the batter. The umpire is behind the catcher. The umpire calls balls and strikes by pointing a finger up in the air. That is to show where he gets his information. Fans sometimes do not agree with the information, which is a reason why beer is served in paper cups.

Fig. 33. The umpire stands behind the catcher and makes the call.

The pitch that goes outside of the strike zone is called a ball even though it already is one. If the pitch goes through the strike zone and the batter lets it go on by, that is a strike. If the batter takes a swing and misses, that is a strike. A batter is allowed three strikes and the pitcher is allowed four balls but not in the usual context. Three strikes and a batter is out. Three outs, and it is the other team's turn to go to bat but it’s like any game; fans don't have to know the rules as long as they can cheer or boo.

The catcher wears a “catcher’s mitt” on one hand and uses the other to point to his private parts. That is supposed to tell the pitcher where to throw the ball, and it is understandable why catchers wear protective gear. Umpires also wear protective gear, mainly to help them with their paranoia.

After a ball is hit, members of the opposing team try to catch it before it hits the ground. If they catch it, it is called a "fly ball," and the batter is out. If the ball is not caught but is thrown to the first base before the batter can get there, the batter is also out. If the batter doesn't touch the base he is out. That really can make it difficult for him to score.

A player scores if he makes it all of the way around the four bases, which are arranged in a square so it is called a diamond. If a batter hits the ball so he can run around all of the bases, that is a home run. Since a home run scores everybody who is on base, some of the fans stand up, clap and spill their beer while others just look morose. From the pitcher's point of view, perfection is when nobody even gets to first base.

The fast ball

A ball coming at 90 miles per hour gives the batter about half a second to make up his mind and swing into action. About half of that time is used up as *reaction time,* the time required for signal a to go from the eyes to the brain where it has to trigger a decision and send a signal all of the way out to feet, elbows, knees and fingers. That means that the batter has to decide what to do when the ball still is roughly 30 ft (9 meters) away and is coming at him like a loose bolt off a motorcycle. Hitting a fast ball is not easy, but meanwhile the pitcher may start to feel like his arm is starting to come unscrewed.

The curve ball

A curve ball is not as fast and is thrown so it spins. Unlike a bullet that spins along the axis of travel, a baseball spins so one side going faster than the other. Air is dragged along and is slightly compressed so it pushes sideways against the ball. On the opposite side suction pulls on the ball, and a standardized stitching increases the drag. A curve ball really does slightly curve, and the amount of curve is emphasized by the point of view of the batter.

The knuckle ball. A knuckle ball is a curve ball with an unpredictable spin so it looks like it is going every which way. A spitball combines features of the knuckleball and the fast ball and is so effective that it is against the rules. It also is why pitchers like to chew tobacco.

The batting average

A batting average represents the chances of getting a base hit. Babe Ruth's lifetime batting average was 0.342, which is near the top and unusual for a power hitter. It means that he had a 34 percent or better than a 1-in-3 chance of getting a hit every time that he went to bat. Ty Cobb had the highest lifetime batting average, 0.366, and he got enough walks that he had a 43 percent chance of getting on base. He also was known for his aggressive behavior. It must have been on his mother's side of the family because she shot and killed his father.

In 1924 Rogers Hornsby's batting average was 0.424, which is the highest on record. He did not smoke or drink so his teammates regarded him as lacking in character.

Joe Jackson , the famous "Shoeless Joe," had a lifetime average of 0.356. He was born in 1887 in South Carolina and never learned to read or write so his wife signed the autographs. He claimed that he got his nickname because in an early game his feet hurt so he took off his shoes, but maybe he was just being a country boy. Joe and seven other players were accused of accepting bribes and throwing the 1919 World Series, and although they were acquitted, the scandal ended his career. His ghost emerged from an Iowa cornfield in the 1989 movie, "Field of Dreams." It was wearing shoes.

Home run hitters

Babe Ruth grew up in a "School for Boys," and was a natural at baseball. It was Babe Ruth who introduced an era of "power hitting" that changed the game. He led a life of plenty with plenty of food, plenty of drink, and plenty of accommodating women. He took better care of baseball than he did of himself. During his later years he would swing the bat so hard that he would almost fall down, but he still could hit home runs. He would dutifully trot around the bases with choppy little steps so it looked like he was having a hard time keeping his front porch out ahead of his vestibule. Throat cancer took Babe at the age of 53. A poignant photo shows Babe Ruth standing in uniform in Yankee Stadium two months before he died, and the photo received a Pulitzer Prize. For all of his triumphs and for all of his flaws, Babe Ruth is baseball

Another power hitter with the Yankees was Lou Gehrig. He was tall, popular, good-looking, and an indefatigable player. He probably would have challenged Babe Ruth's hitting record but he was forced into an early retirement by a tragic and fatal illness now called “Lou Gehrig’s disease. “ He died in 1941 at the age of 37.

What FORE has to say about it

Fig. 34 shows a FORE treatment of the number of home runs hit in major league ball games in the U.S. in a single season. A major shift occurred in 1920 when a harder baseball was introduced that was called “juiced.”

From an average of about 1 home run per game in 1935 the current rate is about 2 per game. Partly this is a result of accommodating designs of some of the ballparks. A separate FORE treatment of the peaks and valleys gives an ultimate average between 2.5 and 2.7 home runs per game, which should be enough to prevent any serious napping in the stands.

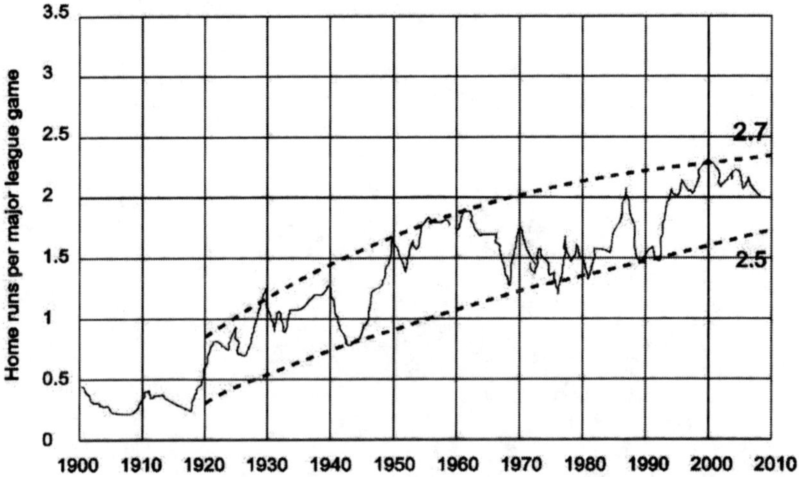

Fig. 34. Home runs per major league game. 1920 marked the introduction of a harder baseball, and the number of home runs doubled. Modified from [http://baseball1.com/statistics/ Sean Lahmann's Base].

FORE and home run hitters

The maximum numbers of home runs hit by one player in one season are shown in Fig. 35, where it will be seen that Babe Ruth was in a class by himself. His records are shown by the open triangles. It took 34 years, from 1927 to 1961, for another ball player to beat Babe Ruth's record of 60 home run hits in a single season. That player was Roger Maris. He hit 61.

A flurry of data points to the right of Babe Ruth's indicates that there was no lack of trying to equal or beat his accomplishment. The emphasis on record home run hitting then faded because trying to kill the ball adversely affects a batting average. Hank Greenberg, Jimmie Foxx, and Hack Wilson came closest to hitting 60 in a season.

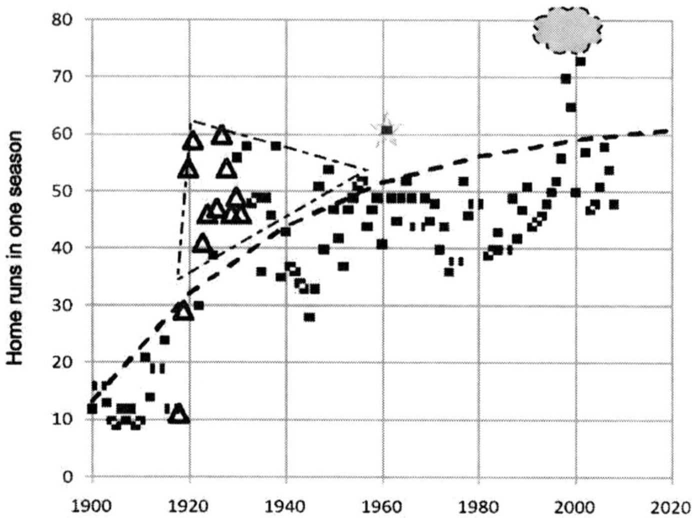

Fig. 35. Maximum numbers of home runs hit by one player in a single season. Babe Ruth's influence is shown in the triangle. A cloud hangs over recent records.

Steroids commonly are used to trick the body into gaining weight, but since most people are trying to lose weight who needs it? Body builders use steroids. Mark McGwire later admitted using performance-enhancing drugs. He said that he did it before it was against the rules, but in order to be on the safe side he kept it secret. Bonds is keeping his secret if there is one. Some of the clues are increasing shirt and shoe sizes, and even now the tests for steroids can be beat.

Table 6. Home runs hit in a single season.

Player	Year	Homers
Rogers Hornsby	1925	39
Babe Ruth	**1923**	**41**
Babe Ruth	**1929**	**46**
Babe Ruth	**1924**	**46**
Joe Dimaggio	1937	46
Babe Ruth	**1926**	**47**
Jimmie Fox	1933	48
Lou Gehrig	1934	49
Lou Gehrig	1936	49
Ralph Kiner	1947	51
Willie Mays	1955	51
Mickey Mantle	1956	52
Babe Ruth	**1920**	**54**
Babe Ruth	**1928**	**54**
Ralph Kiner	1949	54
Hack Wilson	1930	56
Jimie Fox	1932	58
Hank Greenberg	1938	58
Babe Ruth	**1921**	**59**
Babe Ruth	**1927**	**60**
Roger Maris	1961	61
Roger McGwire	*1999*	*65******
Roger McGwire	*1998*	*70******
Barry Bonds	*2001*	*73******

*** Exceed the ultimate projection of FORE.

FORE projects a maximum number of 64 by a single player in a single season. That was destroyed when home run hitting flew off the chart. It eventually got to 73 in a single season. According to FORE, Babe Ruth's 60 was only ahead of his time, but anything over 64 means that there must have been some kind of *major event,* so what was the event? Barry Bonds 73 home runs exceed the FORE maximum by 14 percent.

Big hopes or big dopes

The next question is, should records that involve the use of steroids be disallowed? Lance Armstrong eventually was stripped of his records. Or maybe FORE got it wrong, and 73 home runs by a single player in a season is possible. Anything can be wrong except religion, and that gets a little conflicted.

Height vs. hitting

Babe Ruth did not get his nickname because he was small. When he started playing ball he was too young to sign a contract so he had to be adopted. He was 6 feet 2 inches tall when the average height was 5 ft 7, and he weighed 215 pounds. Barry Bonds is 6 ft 2 inches at 236 pounds, and Mark McGwire is 6 ft 5 inches at 225 pounds, more or less.

Fig. 37 takes an empirical shot at a height factor and looks like a zip code with feathers. Hack Wilson was 5 ft 6 inches and still hit 56 home runs in one season, only four short of Babe Ruth's all-time record. However, Fig. 36 shows Hack's unusual build, a large head and neck, short legs and tiny feet. He still weighed 195 pounds, so any height factor may relate more to body strength that to leg length. If his legs had been stretched out a little he would have been as tall as Babe Ruth. His mother is said to have been regularly smashed before he was born, and as little as three drinks a day can do it. One of baseball's forgotten greats, Hack Wilson died penniless in 1948 at age 48.

Fig. 36. Three of baseball's most powerful power hitters, Hack Wilson, Babe Ruth, Lou Gehrig.

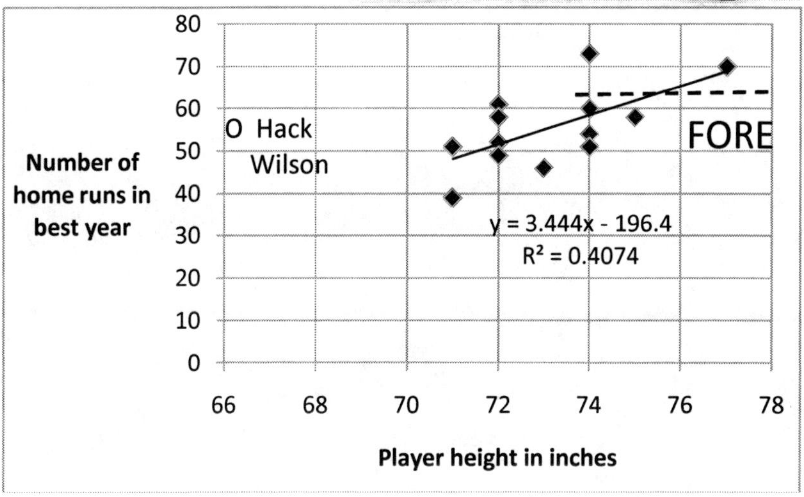

Fig. 37. Height apparently helped Bonds and McGwire go above the dashed line from FORE. An exception was Hack Wilson.

Superstars

A superstar can be defined on the basis of FORE, as a star who is far ahead of his or her time. That even can allow comparisons between stars in different sports. Fig. 37 indicates that Babe Ruth's 60 home runs was 82 years ahead of its time, which may be a record for all sports. Roger Maris was 61 years ahead of his time. As shown in Table 7, Jimmie Foxx and Hank Greenberg also were ahead of their times in terms of home run hitting, as were Ralph Kiner, Lou Gehrig, Joe DiMaggio, Mickey Mantle, Rogers Hornsby and Willie Mays. All are big names in baseball.

Joe DiMaggio had the longest hitting streak in baseball, 56 games. He was 6 ft 1 inch and 193 pounds. His ex-wife was an even bigger hit at 5 ft 6 inches and 120 pounds. She was actress Marilyn Monroe.

Other baseball greats like Bob Feller, Dizzy Dean, Roger Clemens and Whitey Ford rarely if ever hit a home run. They were pitchers.

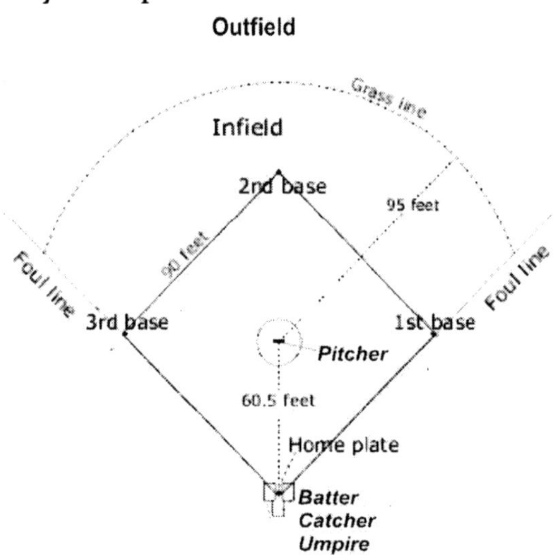

Fig. 38. Baseball field.

Table 7. Years ahead of their times according to FORE, obtained from horizontal distances from data points to the dashed line in Fig. 37.

Player	Year	Homers	Years ahead
Babe Ruth	1927	60	82
Babe Ruth	1921	59	79
Roger Maris	1861	61	61
Jimmie Foxx	1932	58	60
Hank Greenberg	1938	58	54
Babe Ruth	1920	54	50
Hack Wilson	1930	56	50
Babe Ruth	1928	54	41
Babe Ruth	1926	47	21
Ralph Kiner	1949	54	21
Babe Ruth	1924	46	21
Lou Gehrig	1934	49	19
Jimmie Foxx	1933	48	17
Lou Gehrig	1936	49	17
Babe Ruth	1929	46	16
Babe Ruth	1931	31	14
Ralph Kiner	1947	51	12
Babe Ruth	1923	41	11
Joe DiMaggio	1937	46	8
Mickey Mantle	1956	52	6
Rogers Hornsby	1925	39	6
Willie Mays	1955	51	4
Roger McGwire	1998	70	***
Roger McGwire	1999	65	***
Barry Bonds	2001	73	***

*** Exceed the limit of 64 indicated by FORE.

Chapter 6. Will the World Ever Run Out of Oil?

Rag journalism and politics have no appreciation for one-point perspective. They don't get the point.

Tar sands, horizontal drilling, and fracking in the Bakken—this country is on a roll! The first oil well drilled in the U.S. was in 1859 and saved the whales, so one would think that Greenpeace would be more appreciative. Whale oil was the preferred unction for oil lamps until kerosene came along and changed the oil. Whales were delighted.

The common fuel for gas lights was coal gas. "Syngas" stands for synthetic gas but it may also may be a case bad spelling. It was half hydrogen and half carbon monoxide, either of which is capable of sin. If a little air got in a gas line so the light flickered out the choices were to wake up, or die from carbon monoxide asphyxiation, or from the house blowing up.

A memory of Syngas lingers on as toxic residues at former municipal coal gas generating sites. A modern facility still operates in Montana to make hydrogen. It processes the carbon monoxide into carbon dioxide that is piped to Canada to perform bubbly in reluctant oil wells.

Petroleum (Rock oil). In 1901, a well called Spindletop blew the lid off of Texas, and Texas never recovered. Then the automobile gave the oil purpose, and the world never recovered. By 1930 the East Texas oil field made the U.S. the world leader in oil production, and by 1950 the U.S. was producing over 6 million barrels per day. By 1970 it was up to 10 million barrels per day and then, as shown in Fig. 39, U.S. production started to decline as consumption continued to increase. Forty years later production was down by half and consumption had doubled so three-fourths of the oil burned in the U.S. was coming from foreign sources, and every presidential candidate was promising to do something about it. Meanwhile every time the price at the pump went up there is so much wailing and complaining that one might think that the president just died, or that just got re-elected.

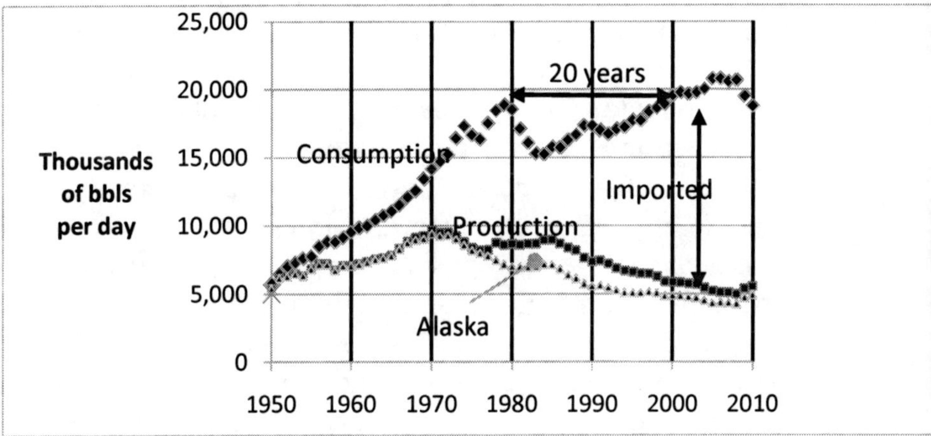

Fig. 39. In 1950 the U.S. went from an oil exporter to an oil importer. A 1970 embargo cut consumption 25% and it took 20 years to recover. Data are from EIA, U.S. Energy Information Administration.

Getting in the spin: The Hughes drill bit

Howard Hughes, Sr., the father, not the famous billionaire son, dropped out of Harvard Law School, then dropped out of the University of Iowa and passed the bar exam anyway. Shuffling papers in a law office was not for him, so he went south to where the action was. He didn't know anything about oil well drilling, but saw there was a problem so he went to work on it.

The problem was that drill bits quickly wore out, after which operations had to be shut down while the entire length of drill rod was pulled up, disconnected and stored in the derrick a section at a time. After the worn bit was replaced, the drill rod would be reassembled and lowered, a section at a time. If during the operation the drill rod got loose, it would drop to the bottom of the hole so there would have to be a "fishing" expedition.

Conventional drill bits had steel edges that scraped at the rock so the edges wore off. The Hughes bit has rollers. It was revolutionary.

The rollers are cone-shaped and mounted at an angle so their tips come together in the middle. That way the cone surfaces roll flat against the bottom of the hole. Hardened steel ridges on the rollers hammer and pulverize the rock instead of trying to scrape their way through it. It was a simple idea that required considerable experimentation. It came at exactly the right time and made a fortune, as virtually every oil well now is drilled with this type of bit.

The fortune was passed on to the now almost legendary Howard, Howard Hughes, Jr.

Fig. 40. An early version of the Hughes well-drilling bit with two roller cones at the bottom. Modern bits are tricone to cut a bit faster. Hughes Tool Company.

The junior Howard Hughes had the flair for publicity. He was tall, rich, good looking, rich, famous, rich, rich, and good looking. He had his father's spark, but aviation was his thing. He designed, built and flew the world's fastest airplanes, almost died in a plane crash, bought Trans World Airlines, and built the largest flying boat in the world that could barely fly. Then he decided to make movies so he bought RKO Pictures, then applied principles of structural engineering to the design of actress Jane Russell's brassiere, which some regard as a forerunner of the geodesic dome. He did pirouettes around tax collectors by giving away billions to establish the Howard Hughes Medical Institute, which became one of the largest and most productive in the world. In his later years, Howard Jr. became a little twisted. He died in 1976, a secretive, sick and lonely man. He did not have a will but did have a secret spouse, and it took 34 years for attorneys to get it all figured out and still have enough left for their comfortable retirements.

Easy oil

Easy oil is never actually easy; it is eas*ier*. Easy oil is concentrated in porous rock layers that hold it like a rigid sponge. Oil floats to the top of water held in the rock pores until it gets stopped where rock layers are broken and shifted by faults, or where they are warped into broad ripples and domes, Fig. 41. A search for oil is not really a search for oil; it is a search for accommodating rock layers. It uses sophisticated tools like echo thumpers and electric resistivity.

Microfossils help to determine geological ages of the rock layers, using fossils that either evolved or went extinct. Over 99 percent of all of the species that once lived on earth are now extinct, which is something to keep in mind when contemplating a war to get even.

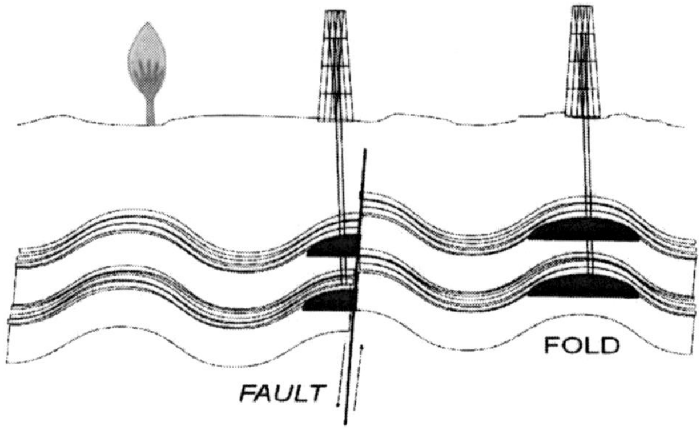

Fig. 41. On the track of easy oil. It floats on water and therefore seeps upward through pores in the rock until it is caught in a "trap." In shale the oil doesn't seep, so it doesn't get concentrated, and it no longer is easy.

Seeping speed.. Oil production requires that the oil must seep into a well as fast as it is being pumped out, which is why the rocker pumps tip back and forth so slowly. If they pump too fast, water pushes past the oil and leaves it behind, and that is the end of the easy oil. That eventually happens anyway, so more oil can be recovered by pumping water down a nearby well to set up an underground circulation and wash out more oil. That is "secondary recovery," but it still leaves about half of the oil in the ground. The world's largest oil petroleum reserves are not just in far-off places but are in old, pumped-out oil fields. What is needed are better ways to get the oil out, but they can't cost too much money.

Fracking

Fracking is not a swear word with an East London accent,

and it is not the same as secondary recovery. It is an old process with a new name. Slow wells used to be punished with a shot of nitroglycerine to loosen things up, but trucks carrying the nitro kept hitting bumps and disappearing. Fracking is safer and works better. It stands for hydraulic fracturing.

Hydraulic fracturing was discovered by mistake in the 1940s. A fluid was being pumped under pressure into the ground, and then the pressure suddenly dropped, indicating some kind of release. An irate farmer then gave a hint regarding what had happened because his field on the other side of the hill was being flooded. Pumping stopped and that could have been the end of it, but if hydraulic pressure opened up cracks in the rock, and if they could be kept open maybe they would increase oil production. They are kept open with fine quartz sand, and

Fig. 42. Spindletop, 1903. It showed that using more straws does not make more oil; after two years production was down over 90 percent. Fracking also does not make any more oil, but helps to get it out.

hydraulic fracturing now is being used for over 90 percent of all onshore gas and oil wells. Now its use also includes getting enough oil and gas out of shale at a low enough cost to jar the economy.

Arabian oil

It must have been part of God's plan for Arabs to get the oil and for Jews to get circumcised. The Arab oil cartel was so upset by the way that Israel conducted the 1967 war, winning it in six days, that in 1973 it punished the U.S. by turning off the oil spigot. However, when oil stopped flowing one way, money stopped flowing the other. The embargo lasted less than a year but the U.S. appetite for oil dropped 20 percent and, as shown in Fig. 42, it took 20 years to recover. Fuel efficiency went center stage as "muscle cars" were wrestled to the mat.

The peak oil hypothesis

In 1956 a geophysicist named M. King Hubbert took his middle name seriously, and predicted that U.S. oil production would peak out around 1970 and then decline. The snickering and knee slapping continued until 1970 when U.S. oil production peaked out and started to decline.

Hubbert assumed that oil production would follow a “normal curve” that is symmetrical and shaped like a bell without the clapper. That would mean that the rate of decline would be a mirror image of the rate of increase, but it isn’t. Hubbert did not have much to go on because the decline had not started yet. A FORE treatment in Fig. 43 is more favorable and projects an ultimate production

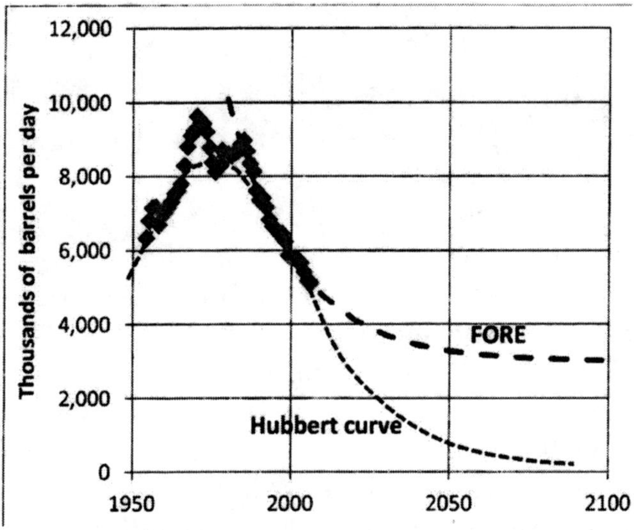

Fig. 43. FORE trumps Hubbert's "peak oil" curve by incorporating effects of price increases on oil production.

in the U.S. of a little over 3 million barrels per day. The reason may be that as production goes down, price goes up and can buy more expensive oil: Oil produced from deeper wells, offshore wells, and deeper offshore wells. A well drilled miles deep does not produce easy oil.

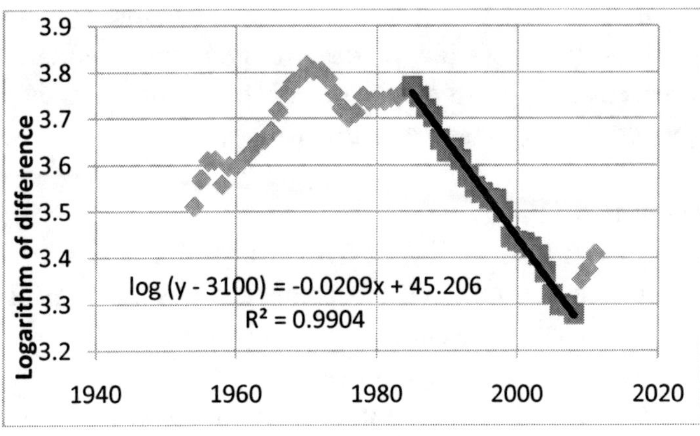

Fig. 44. FORE projected well with a high R^2 value until 2007 when U.S. production suddenly took an

Bringing home the Bakken

The sudden jump in production at the right in Fig. 44 has the look of a *major event.* It happened in North Dakota, which doesn't get all that many major events. Even the movie *Fargo* took place in Minnesota. Now, suddenly, North Dakota is Boom Town, USA, but that happens with every new discovery of oil.

The Bakken shale is not exposed at the ground surface, and was not even named until 1953 when it was found to contain oil. It is about 120 ft (35 m) thick, and even though it underlies a large area, there are limits.

Whereas easy oil is concentrated in rock layers that create traps, shale oil does not seep and therefore is not concentrated. Oil initially forms in clayey sediment, but then gets squeezed out when the fine-grained sediment compresses under its own weight and turns into shale. Shale is by far is the most abundant rock, but most shale contains little oil. It is only the geologically youthful and exuberant shales that have significant oil, shown on a map at the end of this chapter.

Finding the oil in oil shale is easy because it not concentrated in only a few places. It is all over the place. It is the recovery that is not easy. Since it does not seep into pockets, oil wells have to go after it.

Directional drilling

Directional drilling is like mining for oil, as a single vertical boring can spawn off horizontal wells like ribs on an umbrella. Drilling is accomplished by putting the drive motor at the drill instead of at the top of the well, so the drill bit can be tilted and guided electronically. Shale between the horizontal wells is opened up with fracking, an old method and a new ball game.

There are an estimated 4 billion barrels of oil recoverable from the Bakken Shale, which obviously is BIG. Expressed in millions it is not quite as big; it is 4000 million, and the current rate of consumption in the U.S. is about 20 million barrels per day. That being the case, 4000 million barrels would be enough for 200 days.

Let's assume that the amount in the Bakken is underestimated by a factor of 10, the same as for defense spending. That would up the estimate to 2000 days, a little over 5 years. Rag journalism and politics have no appreciation for single-point perspective. They don't get the point.

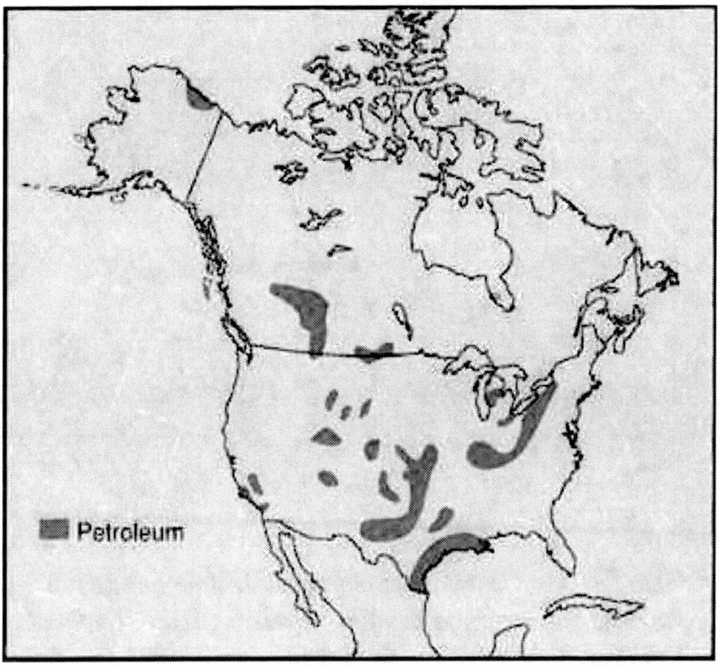

Fig. 45. Where is the oil shale? Most shale is geologically old and so dense that there is little room for oil. Map by the Dept. of Geology, Michigan State University.

How the U.S. was saved by Alaska oil

In the 18th century Alaska had furs, and then the fur trade fizzled out. In the 19th century Alaska had gold, and then the gold fizzled out. In the 20th century Alaska had oil and in the 21st Sarah Palin, but at least now we know what to expect.

After the oil fields in Alaska were heralded as the biggest in North America it was anticipated that they would relieve the dependence of the U.S. on foreign oil, but as shown in Fig. 39, it did not turn out that way. There as a bump in production but it did not keep up with the Arab oil embargo. Then when Alaska oil production peaked out and started to decline the "Star Spangled Banner" started being played over Arctic Wildlife Refuge and the Naval Petroleum Reserve.

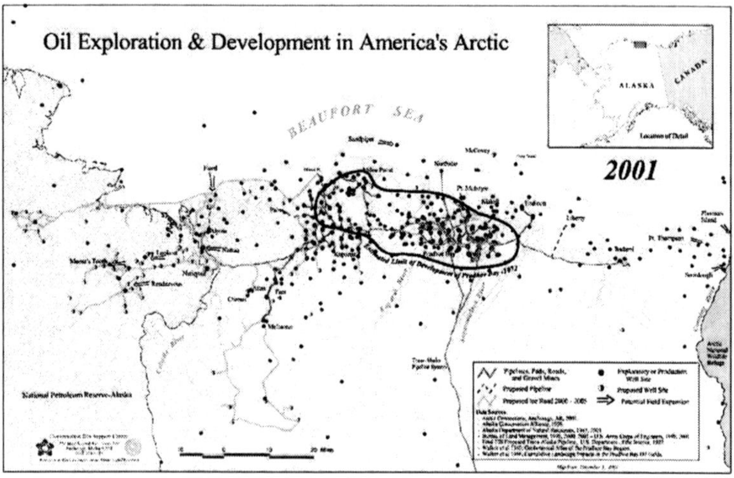

Fig. 46. Map showing exploration drilling on the Alaska North Slope. The Prudhoe Bay field is outlined. U.S. Geological Survey.

Set-asides. The Naval Petroleum Reserve was established

in 1923 in case of war, and various factions have fighting over it ever since. During WWII it was too remote to make a difference, and in 1980 the Arctic Wildlife Refuge was set aside for caribou with a curiosity. It was when the Alaska pipeline stopped flowing full that hungry eyes focused on the reserves. This shows how oil in the ground is better than money in the bank: It gets more interest.

Since the most productive areas already are developed, it is unlikely that opening up the reserves would solve the oil problem. A better prospect is climate change that will open up much larger areas of the Arctic for drilling.

How about those *oil pools?*

"Oil pool" is mainly used to describe investment opportunities because there is no such thing as an underground lake filled with oil. People are misled because pools of water occur in limestone caverns, but that is because the water made the caverns. The only real oil pools are those underneath old cars, although one recently discovered in Nigeria is said to produce high-octane gasoline ready for the tank as soon as the gold particles are filtered out.

A "wildcat well" has about a one-in-ten chance of paying for itself. Occasionally the payoff still is worth the risk, but it is like playing Twenty-one with major oil companies holding the cards.

Hello mudder

One of the most critical requirements for drilling deep holes is mud, the same kind of mud that is smeared on faces to make people think they look younger. Drilling mud is pumped down through the drill pipe all of the way to the bottom of a boring so when it comes up outside of the pipe it brings up drilling cuttings. The rock chips are allowed to settle out in a "mud pit," after which the mud is ready for a return trip.

The pressure exerted by a column of drilling mud that extends hundreds, even thousands of feet into the ground tremendous but it has to be to keep gas pressure in the rock under control. If the mud column is not heavy enough, the well goes out of control and creates a "gusher," which is when people in movies all dance around and get oily because they are rich unless the gusher catches on fire.

In 2010 an automatic shut-off valve on an offshore well in the Gulf of Mexico did not shut off, and the blast killed 11 people. The continuing release of oil turned into the largest oil spill ever, so far. Laurel and Hardy were sent in to do the fix; they put a funnel over the leak to catch the oil and send it up through a pipe to a waiting ship. They neglected to observe that oil does not like going through a pipe or it would not keep trying to get out. There also was another difficulty that would have been obvious if they ever had cooled a beer with a fire extinguisher and then not stopped to drink the beer: Gas released from high pressure expands and gets cold, so cold that methane can freeze water and make the "the ice that burns," and plug up funnels. The well eventually was brought under control the old fashioned way, by drilling an interceptor well in from the side. It should never have happened in the first place.

Gunshots! A completed well is fitted with a large pipe called a "casing." Liquid cement then is pumped down the casing to replace the drilling mud, then come up around the outside and seal off the well. A final step is to shoot the casing full of .45 caliber bullet holes at the producing depths, as a further demonstration of the importance of a liberal education.

Fossil swamp farts

Fracking of shale is producing unprecedented amounts of methane. As costs down, electric power companies are beginning to switch from burning coal to burning gas, reducing carbon dioxide emissions by about one-half and avoiding other complications like acid rain. The coal mining industry is not happy, but it never was. It is a sad situation when a coal miner looks forward to getting black lung so he can retire with benefits.

Smaller amounts of natural gas that are produced from oil wells and refineries often are "flared off," but that is better than releasing it into the atmosphere because methane is a really bad greenhouse gas.

Propane. Propane is manufactured gas, with larger molecules so it can be compressed into a liquid. Propane gas is denser than air so it pools invisibly in low places like basements where lays there waiting for a spark.

In bad odor. Most home explosions occur when people are on vacation so they can't smell the gas. Both natural gas and propane are given an absolutely terrible odor to serve as a warning unless people keep cats. However, a gas leak is not the only way to make a house blow up. A meth lab can do it too.

FORE can't make a projection for the future of natural gas until after production starts to decline. It is declining in Romania, which depends on it for about 40 percent of its energy. However, it is unclear how fracking will play out in the long term for natural gas. An analogy may be to poke holes in a balloon; the gas comes out quickly and then the balloon is empty. After depletion begins, it should become possible to ramp up FORE.

Methane hydrate has been discovered in chunks and nodules in the deep, cold environs of oceans, and there has been idle talk of gathering it up. It may be safer to let the sleeping clathrates lie.

Propane has had an interrupted romance with transportation because of competition from diesel and other fuels. Natural gas would be a natural, but is more difficult because it has to be cooled below -162° C (-260° F) in order to be liquefied and get more bang in the bottle. Both kinds of gas require pressurized tanks and filling arrangements. It is transported on ships in refrigerated pressurized containers.

Fracking, a downside

Pumping sandy water with chemical dispersants into the ground has triggered some worry about possible effects on groundwater supplies, and industry and environmentalists are tooling up for a battle with lawyers working both sides. Vertical wells are sealed off with cement, but horizontal borings do not get that kind of treatment as it would defeat their purpose. Vertical cracks in rock can be very seductive for fracking fluids, but oil wells are deep and water wells are shallow, and hopefully the twain will not meet. The good part is that fracking pressures are only temporary. Pressure has to be removed after wells are completed so the gas can flow the other way.

Earth rumbles. Pumping fluid between rock layers makes it easier for them to slide over one another. The process is called "stick-slip," and when there is a slip it makes an earthquake. The earthquakes usually are small and hardly noticeable, but relieving stress in one place means that the stress will be picked up somewhere else so the next earthquake may be bigger. Fracking pressures are local and temporary, so the effect eventually should go away.

How will shale fracking play out? More holes do not make any more oil, and it is unlikely that things will be much different for shale gas. Secondary recovery methods that wash out more oil may not work as well for gas. Future prospects can be examined with FORE but only if and when production from wells starts to decrease.

Making more oil

Fantasy time—let's make oil out of algae, mold or pollution. Fossil oil got it start as algae growing in shallow seas; it scooped up sun energy and pulled in carbon dioxide to make oily carbohydrates but it did take millions of years of sunshine. The algae being tested to make oil are not just any old algae, but are more like the algae that is used to wrap sushi so it tastes like algae.

Petroleum was produced when masses of dead algae sank to the bottom of seas and got smothered in sediment so there was no oxygen. The algae then were chemically dismantled to make oil and methane gas. Algae farming is on the horizon, but may require a very large horizon to compete with shallow seas that existed during past geological ages.

Uneasy oil

Higher oil prices can pay for more expensive drilling and oil that is not so easy. At first "jack-up" drilling rigs were used in shallow water offshore; they are floated into position and jacked up on legs that extend down into the sea bottom. In deeper water, floating drill platforms are anchored or held in place with propellers, which ramps up the cost. Large underwater pontoons reduce the tendency to rock-and-roll. When a hurricane threatens, everybody hurries home and starts nailing up plywood.

Now Arctic seas that were covered with pack ice are

opening up, and oil companies are trembling with anticipation. Polar bears also are trembling, but for a different reason. They are turning white with worry.

One estimate is that there is enough recoverable oil in the Arctic to last the U.S. for 45 years, if nobody else wants any.

Tar sand rumble

About 25 percent of the oil consumed in the U.S. now comes from Canadian tar sands, where the oil is so thick and gummy that the sand has to be parboiled until tender. The oil floats on water, and sand sinks and is sent back to the pit. Part of the cooking water is re-used and the rest goes in the river. Since the process uses fossil energy to extract fossil energy, it is regarded by some as doubling the downside.

It takes about two tons of sand to get a barrel (42 gallons) of oil. That means that the sand is about 8 percent oil by weight or 22 percent by volume, which is a typical porosity for sand. Venezuela has tar sands but there is too much sulfur, both in the sand and in the politics. Utah has tar sand but it is not negotiable with hot water, which is understandable because they don't even make coffee. If Utah tar sand is heated hot enough, the oil molecules are broken down or "cracked" so they can be gathered up, but a lot of energy goes to heating up the sand.

The tar in tar sand is not real tar. Real tar comes from pine trees and is used for sealing roofs, preserving wood or tar-and-feathering, depending on the circumstances. The so-called tar in tar sand is more like asphalt. It is so gooey that it can't be pumped through a pipeline without a chemical aid. A spill will revert back to asphalt and be hard to clean up.

Pipelines still are safer than shipping by rail, partly

because pipelines can avoid cities while rails seek them out.

Coal oil rumble

During WWII the Nazi war machine suffered from a shortage of oil. At the time Germany was big in chemistry including the chemistry of poison gas. With a proper application of chemistry, coal can be turned into oil. and the peak production in Germany got up to 125,000 barrels per day. That is less than 1 percent of the current rate of consumption in the U.S. . That is when Hitler Youth were sent out on bicycles to stop American tanks but the Americans knew how to handle it. They gave them chocolate.

FORE for oil discoveries

Every new oil discovery means one less left to discover, which looks like it could be a perfect setup for FORE. Discoveries in the U.S. started to decline in the 1930s, and production started to decline in 1970, 40 years later. Oil was discovered in Saudi Arabia in 1933 and production also started to decline about 40 years later. Oil was discovered in the North Sea in the 1960s, and production started to decline 40 years later. According to that timing, Bakken production should start to decline around 2050. The life of a conventional oil field is extended by secondary recovery, but that may not work in shale.

According to Fig. 47, new oil discoveries worldwide started to decline after 1970. The downhill slope is jagged because discoveries tend to be jagged. Statisticians use a "moving average," but as usual FORE marches to a different drummer. When FORE is applied separately to

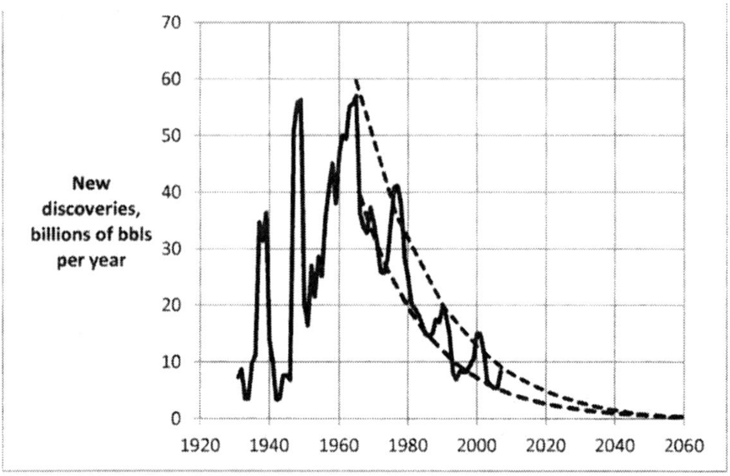

Fig. 47. New world oil discoveries are in a decline and if the trend continues will zero out around 2060. Data are from oil-discoveries.jpg, quoting from Exxon-Mobile (2002).

peaks and valleys in discovery rates the two lines merge and give the same projection. Production from the last major discoveries should peak around 2060+40 = 3000.

Energy alternatives

Hydrogen action. From environmental considerations hydrogen is the perfect fuel. It burns to make water: 2H + O = H_2O. Natural gas or methane is only half as perfect because it burns to make carbon dioxide plus water. Coal is least perfect; it only makes carbon dioxide.

However, hydrogen is not a fossil fuel but has to be manufactured, and at best the energy required to make it equals the energy that is released when it burns. That makes it like a storage battery.

Electric action. Hybrid cars have storage batteries that are kept charged with on-board generators. They run on petrol, and only as needed. Braking recharges the battery, but unfortunately it is not possible to get anywhere by braking all of the time. The weight of the batteries is partly compensated because there are no clutch or transmission. The same principle has been running trains for decades, ever since demise of the steam engine.

Moonshine over the mountain. Alcohol is fuel. It thins paint and thickens the brain. When added to gasoline it prevents engine "knock," where the other alternative is a lead compound that goes out of the car tail pipe and out into the food chain.

In the U.S., alcohol for fuel comes mainly from corn, which requires fertilizer, insect control, plowing and harvesting. Each one takes another swipe at the environmental benefits. A better option is to make alcohol from cellulose—cornstalks, wood chips, paper, cardboard, etc. Switchgrass also works and is perennial, native and requires no fertilizer. It also restores the soil, so it may have a lot of things going for it.

Getting alcohol from cellulose emulates digestive functions of termites, with hints obtained from panda poop since they barf down bamboo shoots like celery. First the cellulose goes to sugar, and then fermentation turns the sugar into alcohol. A FORE projection for cost to make alcohol from cellulose is shown in Fig. 48 and already is out of date. The symptoms are those of a *major event* because there has been almost a 40 percent reduction in cost in four years. Production may become cost-competitive to gasoline by 2016, which is catching the attention of the petroleum industry. However, switchgrass at best it will be a minor player because land that is used to grow switchgrass is taken out of food production.

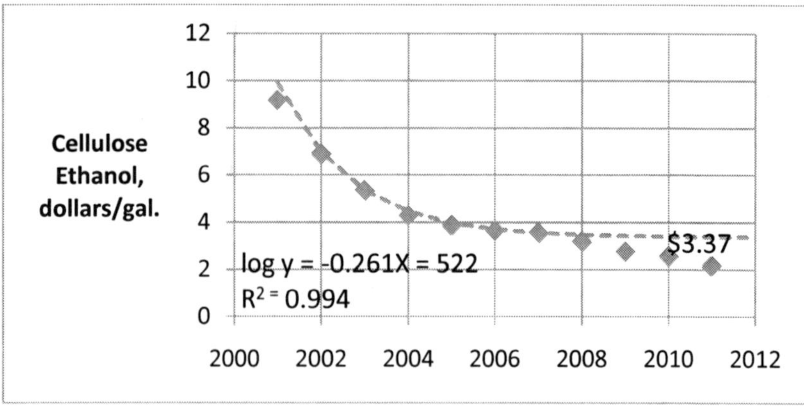

Fig. 48. The cost for producing alcohol from inedible cellulose was following FORE but now may be entering a new phase.

The wrong end of the cornucopia

Transportation facilities have a cornucopia of funding but the cornucopia is open at the wrong end. Railroads build their own tracks as a penalty for their efficiency, but highways and bridges for the most part depend on "use taxes" that are indexed to amounts of fuel being used. From 1950-1980 fuel consumption in the U.S. went up by a factor of four and helped to pay for the Interstate Highway System; build it and they will come, and they came. The system now is over 50 years old and ready for an overhaul, but now the emphasis is on getting more miles to the gallon. Fewer gallons, less tax money but meanwhile repair and replacement costs are indexed to inflation. That is no way to run a store. A clever option is the toll road where users pay twice, once for the gas tax and once for toll. That's how to run a store.

Thoughts for the long run

• Will coal be able to survive the natural gas war anywhere except in China?

• If burning natural gas reduces carbon dioxide emissions by half, what about the other half?

• The rest of the world still is learning about cars, so what happens when it latches on to air conditioning?

• How do you fly a kite when there isn't any wind?

• How does one get electricity out of a reservoir that doesn't have any water?

• What can be done with a reservoir that is full of mud?

• What will happen when the costs and human misery caused by climate change exceed the costs of trying to fix it and it is too late to fix it?

• If geophysicists are right that most of the earth's internal heat results from radioactive decay, then geothermal energy is ultimately nuclear, from deep within the earth; does that mean we can't use it?

• If the 20th Century turns out to be "The Oil Century," what will be the 21st? Dark ages?

• Oh, hell; let's get on to the next chapter.

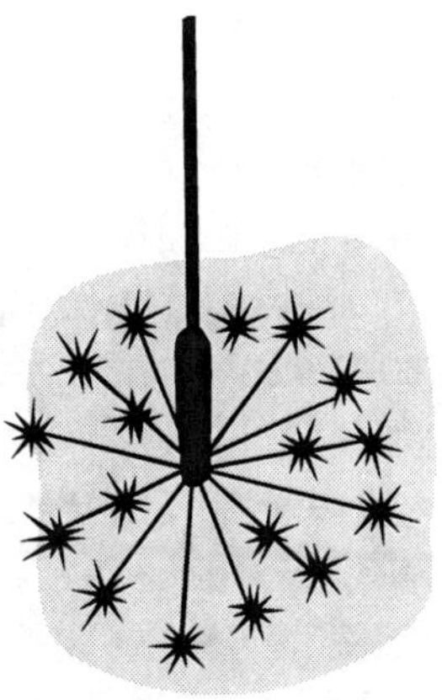

Fig. 49. Directional drilling and fracking are like sparkle plenty.

Chapter 7. Womb to Tomb

Human brains consume a lot of calories, but as far as we can tell nobody has been trying very hard to lose weight by thinking.

The most troubling aspect of life expectancy is because it is called life expectancy. Life expectancy does not mean how long a person can expect to live. Life expectancy is a statistic. Mark Twain laid it on that one when he observed that there are three degrees of prevarication: lies, damned lies, and statistics.

A life expectancy is automatically awarded to every new person at the instant when he or she gets pushed out into the open when what they want is some fresh air. Life expectancy in the U.S. currently is about 78 years, more or less, more for women and less for men. However, the table show that when a person lives out his or her life expectancy, he or she suddenly gets awarded 10 more years! That is, at 78, the life expectancy becomes 88, simply as a reward for living. So how cool is that!

Then at age 88 there is another award, 5 more years, plenty of time to go ahead with the breast implants and begin live a little.

Another payoff occurs at age 93, but the bonuses are diminishing so one probably should try and avoid getting any unplanned pregnancies or engaging in any other long-term commitments, and there comes a time when one should not buy any more green bananas. It nevertheless appears that the longer a person lives, the longer a person lives. There is an element of poetry in that.

The Catch of the day. Life expectancy actually is death expectancy made politically correct. Life expectancy means the age at which one-half of one's contemporaries will be dead. That is a sobering thought. It means that there is only a 50 percent chance of living out a life expectancy. Life expectancy is more a death expectancy, and there is no getting around it. There is a 50 percent chance that a marriage will end in divorce, but nobody calls it "marriage expectancy." It is divorce expectancy. Let's get real.

Here comes life expectancy. The person who makes it past their first life expectancy then gets a 50 percent chance of making it past the second. Fifty percent of 50 percent is 25 percent. It means that the barely acceptable 1-in-2 chance of making it to age 78 drops down to 1-in-4 of making it to 88 and meanwhile one is going to lose a lot of contemporaries. There is a 1-in-8 chance of making it to 93 so there is likely to be a small funeral with or without the usual fanfare and lunch featuring potato chips and a green-bean casserole. As one gets older the transition can be easier.

A normal distribution

A "normal distribution" is another statistical stretch, the curve that is shaped like a bell without a clapper, but in this case it must have been a bad day at the foundry. A normal distribution is supposed to be symmetrical, which would mean that for every baby that has the misfortune to die at birth, another person should make it to 2 × 78 = 156 years old, and that has yet to happen. That is because the life expectancy curve is "skewed," another case of bad spelling. Then to compound the felony, in statistics an average is called a "mean."

Another term that is highly questionable is called the "standard deviation." It is supposed to represent the spread or variability of data, but most people do not pony up to the idea that a deviation is standardized, particularly if it is seen hanging around the schools.

Historical death expectancies

During the Ice Ages, "animal rights" meant the right to eat animals, which could go either way. The concentration of calories in meat are supposed to be what allowed humans to evolve large brains and become more thoughtful. Our large brains consume a lot of calories, but as far as we can tell nobody has been trying very hard to lose weight by thinking.

The life (death) expectancy of early humans is written in their bones. People usually did not live long enough to get arthritis. For thousands of years the average life expectance hovered around 35 to 40. Surprisingly, as shown in Fig. 50, that relatively low life expectancy lasted all of the way into the 1800s. Then life expectancy took a jump and started to level out in a manner prescribed by FORE, shown by dashed lines in the figure. Indications are that it will only take about 190 years for the life expectancies for men and for women to be the same. We can hardly wait.

The data used in the figure are for whites in the U.S., not because of racism but because the data are more consistent. It may have to do with living in a lead-free neighborhood.

The linear FORE relationships for Fig. 50 are shown in Fig. 51 and again indicate that men and women really are different , even though the eventual average checkout time for both comes out the same, around 86 years with a 50 percent chance of living longer. When Social Security; was established in the U.S. the average life expectancy was lower than the retirement age of 65, and Social Security made money. Now that is changed.

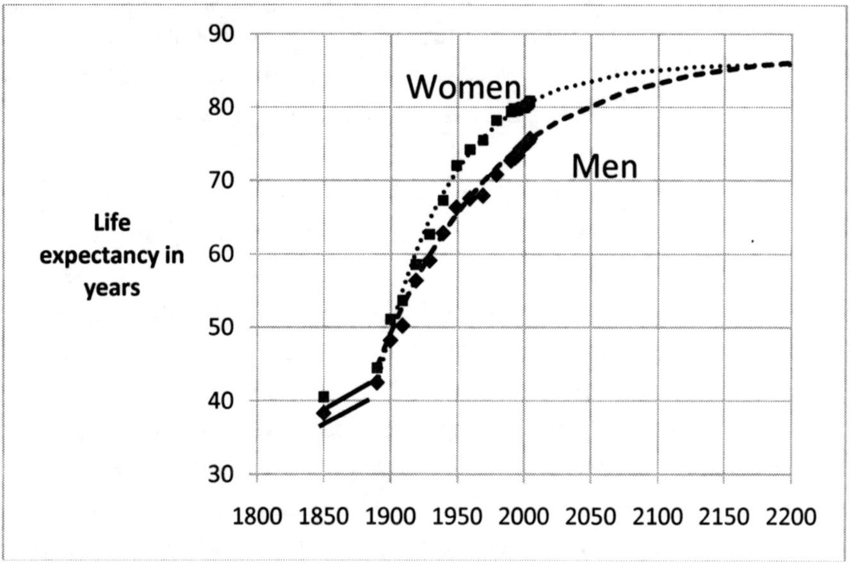

Fig. 50. Women may not always live longer than men if men can calm down and quit smoking. The two FORE projections come out about even.

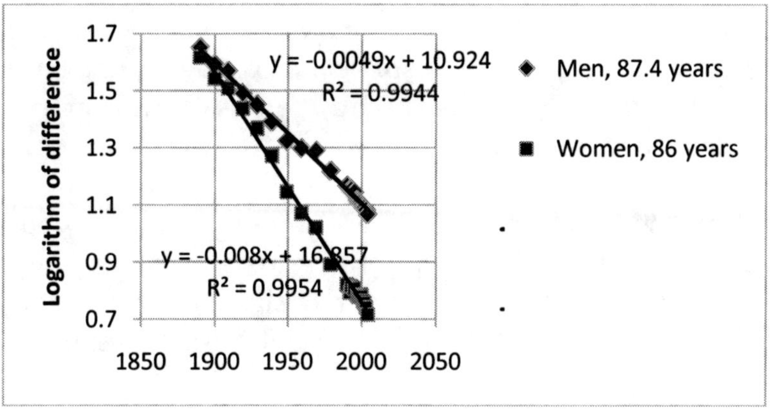

Fig. 51. The linear FORE relationships used to draw the lines in Fig. 50.

ATTENTION! FORE shifts gears in this chapter because numbers increase instead of decrease with time. However, the *rates* of change still decrease so FORE still works. All that is required is to change + and - signs, and there is an automatic warning because there is no logarithm for a negative number. FORE depends on and defines an ultimate value whether it is goes up or goes down.

A brief history of life expectancy.

Sir Walter Raleigh introduced tobacco to England, which could have been a blow to life expectancy except that it came at a time when people didn't live very long anyway. He named Virginia after Queen Elizabeth I, and then attempted to make some extra points by laying his coat down in a gutter so the Queen could walk across without getting her feet muddy. Unfortunately since the coat was over an open manhole he got the usual one-way ticket to the Tower, where he stopped smoking after his head became disconnected and he stopped breathing. It can be extremely difficult to break the tobacco habit but the method works.

A major event. In Fig. 50 the sudden jump in life expectancy after 1800 has the look of a *major event,* so what made the event? In 1850 the life expectancy of a boy at birth was only 38 years, and many sad little grave markers only say "Baby" so they could save the name for the next one.

Smallpox took the lives of about 1-in-3 of those who were infected, and left the other two scarred for life. Where the Black Death required the cooperation of infected fleas, smallpox only needed a shared spoon, blanket or a sneeze.

In the late 1700s an English surgeon named Edward Jenner made his famous, world-class empirical observation regarding milkmaids, so he was not just looking at their breasts. He observed that if the milkmaids got a relatively mild disease called cowpox, they did not get smallpox. He therefore tried inoculating people with cowpox, which was high risk because if something had gone wrong he probably could have been called a witch, or the very least would have been shipped off to Australia. The process worked, came to be called "vaccination," *Vacca* being Latin for cow.

Smallpox vaccinations were one of the greatest triumphs of modern medicine prior to invention of Viagra, and shows what can be accomplished through simple observations. The most virulent killer disease in history now is gone except for a few samples kept locked away in secret places.

Vaccinations then were developed to combat other fatal distractions like diphtheria, mumps, measles, lockjaw and whooping cough. Every time a new disease or a new version of a flu virus comes along, the search is on for a new and better vaccine.

Pasteurization. Louis Pasteur was a French scientist who invented polarized light, proposed germ theory, and developed a way to store milk and beer at a time when there was no refrigeration. The latter process is named after him, "pasteurization." It uses moderate heat and does not actually kill all bacteria but stifles their enthusiasm.

Pasteur went on to devise a vaccine that can prevent rabies, commonly called "lockjaw." Rabies was transferred by animal bites and inevitably was fatal. Pasteur tried his vaccine on a young boy and even though it worked, they still got him for practicing medicine without a license. Professionals always have to be on their guard.

Fig. 52 shows dates of discovery of various vaccines. Most came after 1920, too late and too spread out in time to qualify as the FORE major event.

How to stay safe without being vaccinated

Since vaccinations are regarded by some people as invasions of personal freedom and tinkering with God's will, they don't get vaccinated and don't have their children vaccinated, yet they still do not get the killer diseases. That is God's way, to put a cow in a herd that has been vaccinated, as it is impossible to get a disease if there is no disease there to get. The formal name is "cocoon effect." It is an important aid to self-righteousness but they and their children better not make any trips to places like Calcutta.

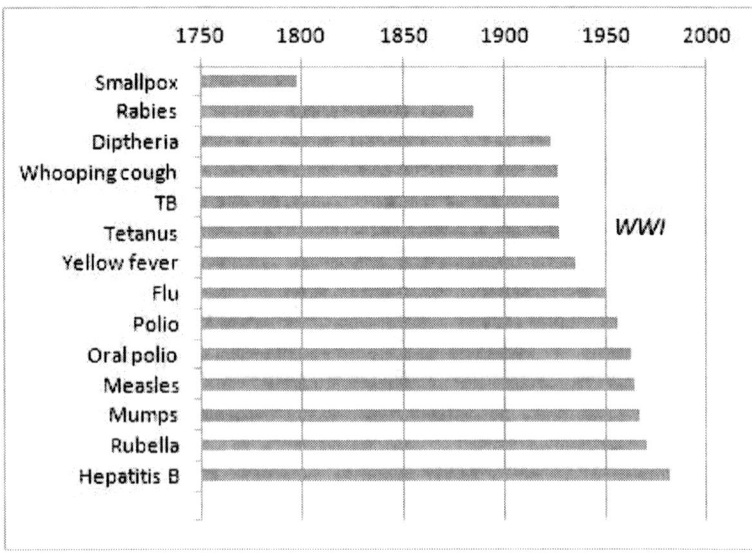

Fig. 52. Historical discoveries of new vaccines. Most were after 1920.

Antiseptics and anesthetics.

Surgeons once regarded pain as reassuring because it meant that that a patient was not dead. One way to ease the suffering was with a little medicinal alcohol but then the doctor sometimes forgot where he left his tools and had to go back in after them.

It was dentists who discovered that pain was not good for business, and in 1844 a dentist named Horace Wells tried easing the pain with nitrous oxide, or laughing gas. That helped but it was not easy to work on a patient who was laughing his head off. Surgeons then tried giving patients ether and telling them to take a deep breath, but it is questionable how much that might have affected life expectancy.

The royal flush

One of the less celebrated thrones in the Hampton Court Palace is a wooden seat with a hole in it, where King Henry VIII is said to have sat while contemplating his movements. Sanitary facilities were relatively uncomplicated in those days, next to a moat.

The flush toilet commonly is attributed to an English engineer named Thomas Crapper, but Roman engineers were building crappers long before there crappers were named in the honor of old Tom Crapper. A Roman crapper is shown at the end of this chapter. It is a long stone seat with holes in it, and there was water flowing past under the seats to kept movements moving. An adaptation without the seats often is pay-as-you-go as an attendant to keep track of the assets. Remember, you saw it here.

Rome had bragging rights for its *Cloaca Maxima,* which is Latin for "Sewera Biggesta." The Sewera Biggesta dumped into the Rivera Tibera, creating a precedenta that continues to this daya. People downstream can tell when asparagus is in season.

Fig. 53. 2000-year old throne room in Ephesus, Turkey, where Romans sat and did their business after some bare-assed slaves had warmed up the seats. Water flowed past under the seats to keep movements moving. It looks like some-body has a regrettable weight problem. Author photo.

City life. In medieval cities, chamber pots were conveniently dumped out of second-story windows, which probably led to the customary, "Have a good day!" That also is when it was determined that men should walk on the outside and women should wear large hats with streamers.

The most common formal facility was the “privy,” which meant privacy with an attitude. If the privy was too close to a well there could be "pump-a-dump," which was terribly unhealthy. After cities ran out of places to put privies the option was sewers where the most convenient outlet was into a river. In 1850 the Thames was so polluted that lime-soaked drapes were hung over windows of Parliament to try and kill the odor. So many children died from cholera and typhoid that people were encouraged to have large families in order to maintain a tax base.

Early sewers were made with bricks and leaked. Even today sewer pipes can leak, usually at joints or through cracks caused by weight of the overlying fill soil. Leaks can travel a long way through loose soil around a buried pipe, so a particularly bad situation is where a sewer pipe and a water pipe are close to each other or are laid in the same ditch. If and when water pressure ever goes down, the leak into the water pipe and can cause bombastic bowels and potentially fatal diseases like cholera, typhoid and hepatitis.

Chlorination and improved sanitation. Chlorine is part of common table salt, sodium chloride. Divorced from sodium it is a poisonous gas, but still is used in small amounts to disinfect swimming pools.

In 1908 chlorine was added to water supplies in Jersey City, and there was a 90 percent reduction in deaths from typhoid fever. By the 1920s nearly all of public water supplies in the U.S. were chlorinated without much argument. Improved sanitation coincides with the sudden surge in life expectancy. To prove the point, life expectancy remains low in areas where sanitation is poor and survival depends on "natural immunity" that can result from prolonged exposure among the survivors. Sanitary engineers, now called environmental engineers, indirectly have saved more human lives than have the medical professions. Safe drinking water remains the most important of all preventative medicines.

Smoked lung

In 1900 less than 1 percent of the U.S. population smoked cigarettes. Men did not smoke cigarettes because cigarettes were not masculine, so they chewed and spit or smoked cigars and polluted the atmosphere. Cigarette smoking got a boost from movies showing beautiful people blowing smoke at each other in a highly suggestive manner. Now movies skip the preliminaries and get right down to business.

During WWII cigarettes were distributed free to service personnel, and after the war tobacco companies extended their generosity to include students on college campuses. It was not easy to turn down free cigarettes, and by the mid-1960s a majority of men and a third of the women in the U.S. smoked cigarettes. It was the "in-thing" to puff on an adult binky, squint and tilt the head to let the smoke come out of an ear. Once a person was hooked, it was about as easy to stop smoking as it was to stop breathing.

The unsecret ingredient. An average cigarette contains about 1 milligram of nicotine, and filters don't take it out. Nicotine is a highly addictive, habit-forming stimulant. For years nicotine was extracted from tobacco to make a very effective ant poison, until it was considered too risky. Lungs are well adapted for a quick delivery.

When it was discovered that smokers put non-smokers including children at risk, education, regulations and vigorous campaigning started to convince smokers that they had been duped. Cigarette companies had to adjust and they pushed foreign sales. Meanwhile, as shown in Fig. 54 smoking rates in the U.S. are down by over half, and the declines are going according to FORE. The ultimate projections are about 11 percent of men who still smoke and 4 percent for women because of their greater won't power.

All habit-forming drugs from meth to marihuana work by bringing out the dope in a person's dopamine while shutting down brain receptors so they keep asking for more, but only for medicinal purposes.

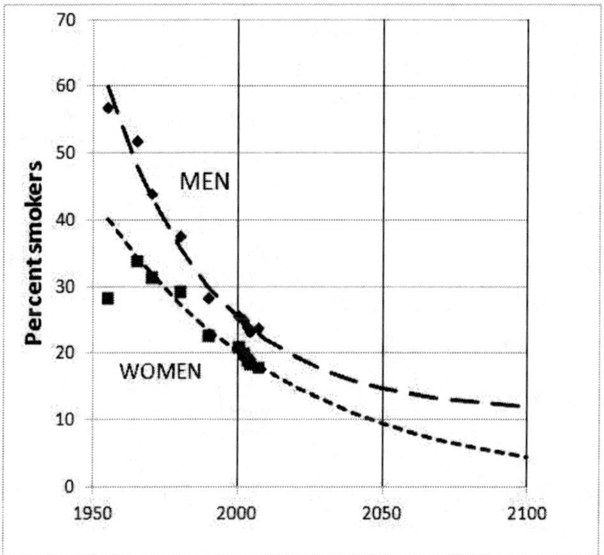

Fig. 54. FORE and tobacco smoking in the U.S., where women must have greater won't power.

The fat in fatality

A hundred years ago people came into Ellis Island hungry, and they have been overcompensating ever since. Obesity now is defined on the basis of weight and height measurements: A person whose weight is more than 20% over his or her hypothetical ideal weight is said to be obese. The definition is attributed to a committee of thin people so the easiest way to conquer obesity is get some fat people on the committee.

Fig. 55 shows the U.S. Center for Disease Control take on the problem. It is a straight-line increase that definitely is not FORE. Colorado is the least obese state in the U.S. but that is because it takes 20 minutes to boil an egg. At the other extreme are Alabama, Mississippi and Tennessee, which confirms the superiority of southern cooking.

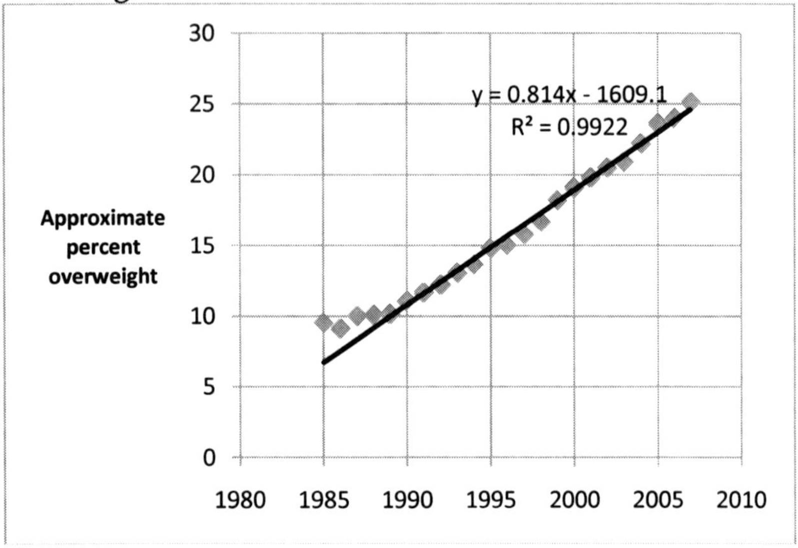

Fig. 55. No leveling off and no food for FORE, but it can be expected to will level off because of dieting or because the percentage can't exceed 100, whichever comes first.

A classic experiment. Two groups of people, a fat group and a thin group, were given all of the ice cream that they cared to eat, and there were two kinds of ice cream. One kind was thick and creamy and tasted good and the other probably had freezer burn. Result: The chubbies ate more of the good ice cream and less of the bad, and the skinnies ate about the same amounts of each. An obvious way to reduce obesity is skip the chips and pass the rutabaga. The people eat too much because they like to eat. It may be inherited from earlier days when food spoiled if one did not eat it. Simba gorges after a kill, then sleeps it off.

Fig. 56. Famous actress Lillie Langtree who was renowned for her beauty and did not let her talents go to waist. She was born in England in 1853 and died in 1929 in Separation, Utah. It figures.

Fig. 57. We would, but we're

- *too young.*
- *too old.*
- *too busy.*
- *too tired.*
- *too sleepy.*
- *too small*
- *too big.*
- *too fat.*
- *out of here.*

www.livehealthmedia.com.

Chapter 8. A Growing Population

In ancient Egypt birth control involved the use of crocodile dung, which must have been a real downer for the men.

In 1798 an economist named Thomas Malthus proposed what now is called the Malthusian Doctrine. A doctrine is a doctrine and it requires no scientific proof, but a little proof is not objectionable.

Malthus' argument was that if two of anything reproduce and make four, and the four reproduce and make eight, and so on, a population will double with each succeeding generation. Here time is on our side, because according to the doctrine of Adam and Eve there have not been that many generations. If Adam scored with Eve 4000 years ago and a generation averages 25 years, then $4000 \div 25 = 160$ generations. Population dynamics being what they are, that would only create 2^{160} people, which is 2×2 one hundred and sixty times. Let's try that out: $2 \times 2 = 4$; $2 \times 4 = 8$, and so on. And the answer is:

1,461,000,000,000,000,000,000,000,000,000,000,000,000,000,000,000,000,000

give or take a few. It therefore appears that Malthus' argument needs some restructuring since the world population is only around 7,000,000,000, hardly enough to worry about.

According to Fig. 58 the world population did double between 1960 and 2000, but it took 40 years instead of 25. Even though 40 years appears to be pushing the child-bearing age, it is hard to argue with statistics. That reduces the number of generations down so 4000 ÷ 40 = 100 instead of 160. The revised Malthus calculation gives.

1,267,000,000,000,000,000,000,000,000,000

still too many. Maybe Malthus had been drinking.

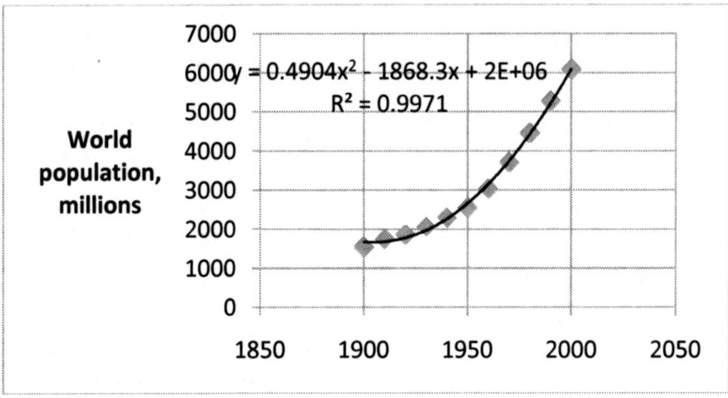

Fig. 58. The world population doubled between 1960 and 2000 with no sigh of leveling off and no FORE. This is not a good sign.

The forever food supply. Malthus proposed that food supplies also will increase, but not as fast as populations. He suggested that as populations go according to 2, 4, 8, 16, and so on, food supplies only go as 1, 2, 3, 4, which makes famines inevitable unless something can be done to outsmart them. Famines occurred in the Soviet Union in 1921, in China in 1959-61, and in North Korea in 1996. Coincidentally all were being managed under Communism.

Malthus' model is shown on the left in Fig. 59. In the right graph it is modified to be in better agreement with observations but consequences remain the same.

Managing motherhood

In ancient Egypt birth control involved the use of crocodile dung, which must have been a real downer, especially for men. Around 600 B.C. a natural herb called *silphion* was found to be an oral contraceptive and was harvested to extinction. Casanova's tool case is said to have been made out of sheep gut but it is unclear if it might still have been part of the sheep.

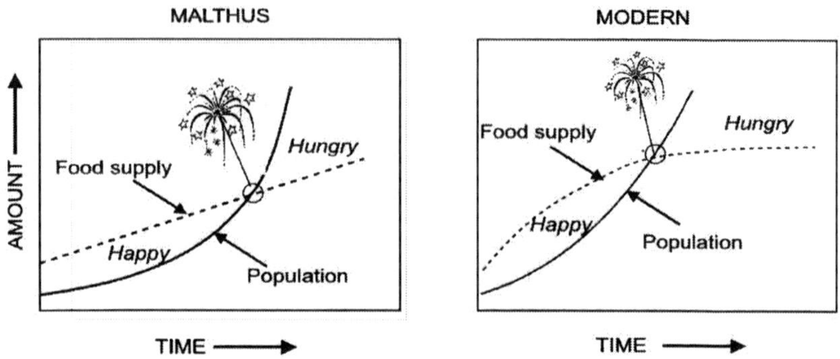

Fig. 59. Malthus got the idea across but his doctrine could use some modification.

Starting in the mid 1800s, vulcanized rubber was discovered to give greater mileage and better traction, and worked so well that in 1873 the U.S. passed a law making contraception illegal, after which people had to be more careful about pulling down the shades. The Supreme Court eventually struck down the law although the members themselves probably were indifferent.

The "rhythm method" got a push in the 1930s and then pushed back when it started to be referred to as “denial days." A problem is that if a person wakes up in the middle of the night and can’t remember his phone number he may not know what day it is.

Then the pill came along and changed everything. It might even work for animals so instead of hitting a deer one could stop the car, jump out and say, "Here, take this pill."

Taming the population numbers

Troublesome trends can be made more acceptable by simply using a logarithmic scale because it shrinks them down, shown in Fig. 63. Then a straight line only looks linear because what it really means is that growth is *exponential*. Instead of increasing according to x it is increasing according to 10^x, ' but it still looks better. Also the graph puts the burden on time instead of on family size, so it relieves individuals of any responsibility.

Around 1950 the *exponential* rate of population growth took a jump, which is *really* bad news regardless of responsibility. What happened?

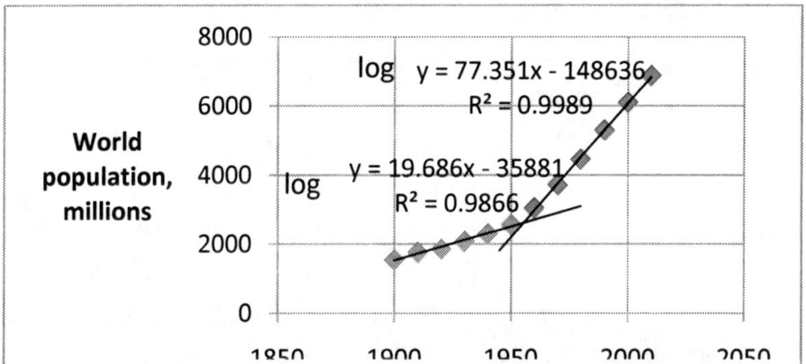

Fig. 60. The logarithmic scale is not good news. It supports Malthus' hypothesis instead of FORE.

Wars and depressions. Populations always take a jump after a major war. Maybe it is instinctive, to try and make up for the losses. The opposite occurs during a depression when an immediate concern is the number of mouths to feed, but nobody wants a depression. Between times programs like Save the Children might achieve a better balance if they included some information on how not to have so many children.

California concept

California is North America's bumper as the entire continent keeps moving west and trying to push it off into the sea. California keeps humping and bumping along, overriding a submerged mass of rock called the Pacific Plate and creating deep scraping noises that constitute earthquakes. As Pacific air is pushed up into the hills it dumps rain that promotes landslides. However, the mountains also form a barrier between warm Pacific air cooler Canadian air masses, which puts a stop to tornadoes. Eastern states in the U.S. get nor'easters, central states get tornadoes, southern states get hurricanes while western states have to settle for deserts, Santa Anna winds and forest fires.

Those are only a few of the things that make California so interesting.

Westward, ha! The center of population in the U.S. started to move west after the discovery of gold in California, and south after the discovery of air conditioning. Since California was the last stop before the ocean it was almost inevitable that it would become the most populous state in the U.S.

Fig. 61 indicates that population growth in California may be following the Malthusian model, in which case how long can it continue? Will California run out of space, water, money, or enthusiasm, or all four hit at the same time? Will California end up like New York City, with no place to park? The treasure is in the details.

Details. In 1848 California was snatched away from Mexico, and Mexico has been trying to get even ever since. According to Fig. 62, without the Hispanic immigrants the population of California would be leveling off. It would be following FORE toward an ultimate target of around 28 million. Instead it is already pushing 38 million and the growth rate shows no signs of easing off. Contrary to some popular opinion that is *not* un-American; it is American. It has been nonstop ever since the pilgrims got off the Mayflower.

What if the influx of new immigration were to level off? It looks like the rate of population growth of California also would level off. What about immigrants having large families? That may ease back when they see how much it costs to send the kids to college so they won't have to spend all
of their lives picking somebody else's grapes.

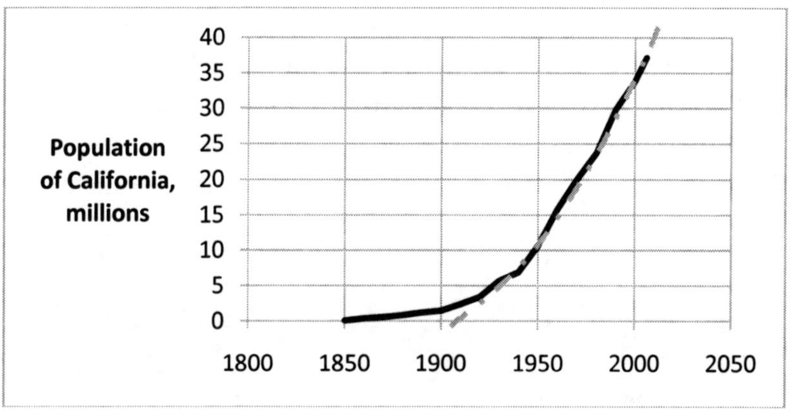

Fig. 61. Population growth in California. The dashed line shows the Malthus relationship. This is not good.

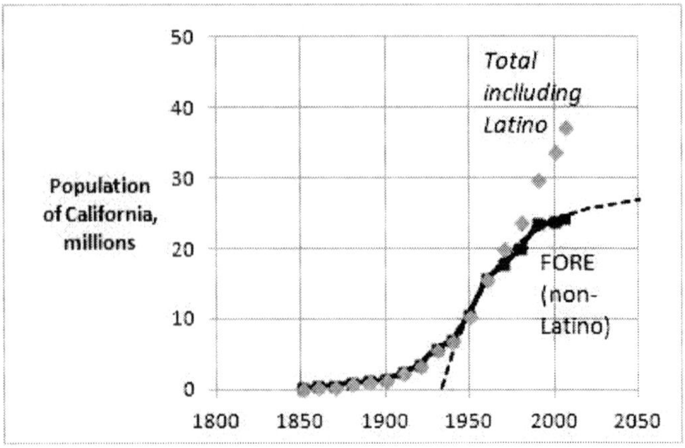

Fig. 62. The population growth in California follows FORE if recent immigrants are filtered out.

What's next for California? The baseless speculation in Fig. 63 was obtained by sliding FORE up a notch, based on two assumptions: (1) immigration will ease off, and (2) assimilation will lead to smaller families. The projection then is for an ultimate population of about 50 million, of which 50 percent will have a Latino heritage but all will be Americans. That is a guess with just a dash of sophistication.

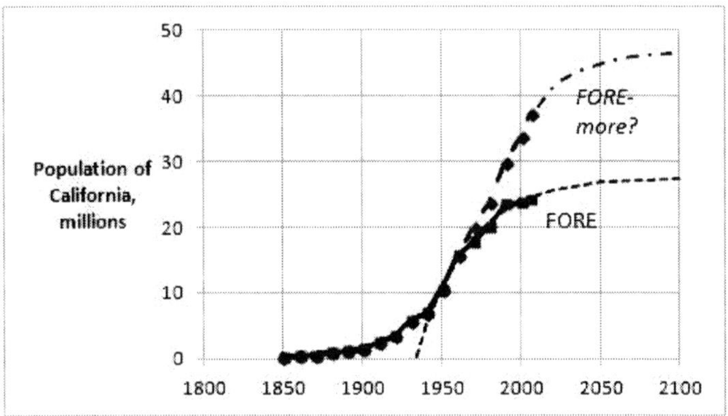

Fig. 63. If immigration and baby manufacture in California ease down.

How it is in the U.S.

The population of the U.S. also has a Latino component, Fig. 64, but the population exclusive of Latino's appears to be entering an early stage of FORE. The FORE projection without Latinos indicates an eventual population increase of about 50 percent, to 450 million. Adding an arbitrary percent for Latinos would run that up to 540 million and more Mexican restaurants.

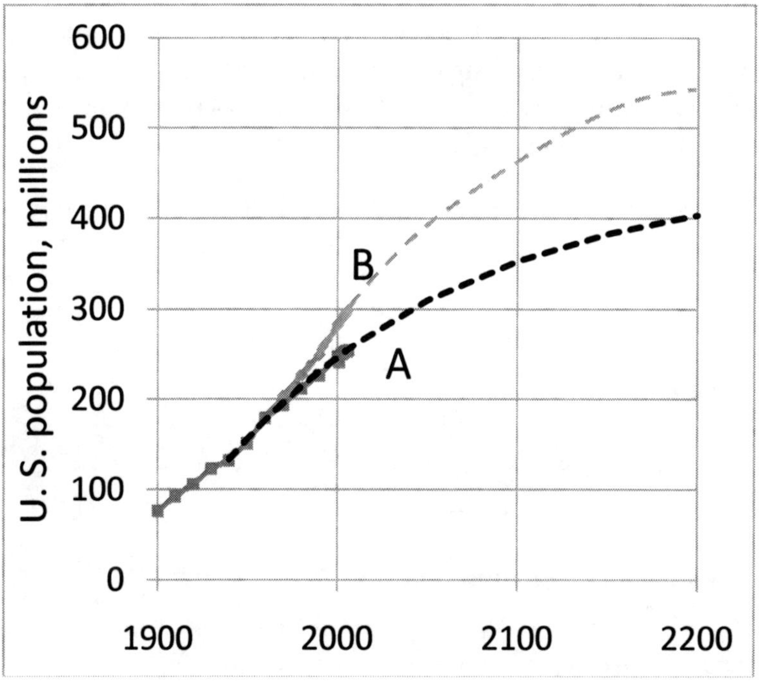

Fig. 64. Maybe this will happen If both immigration (the difference between A and B) and baby manufacturing eases down.

A maturity factor

In the early days of settlement Iowa meant farming and it still does rather well at it. As transportation improved and farming gradually became industrialized, rural populations dwindled. The transition to city life was not without an element of nostalgia, re: Grant Wood's "American Gothic." Now a single farm can cost more than Park Avenue penthouse and likely be more productive. Iowa has about one-third of the area of California and one-tenth of the population, and has a consistently low unemployment rate.

As shown in Fig. 65, after an initial surge the population of Iowa started leveling off and following FORE even with contributions from immigration. Changes have been swift, and the state has matured.

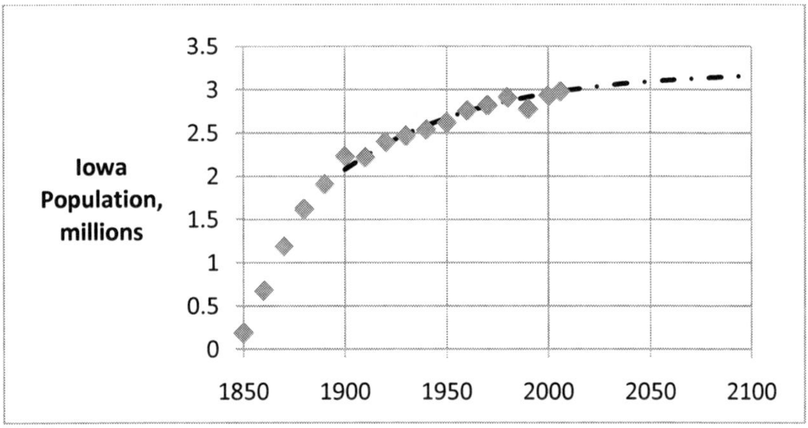

Fig. 65. After a burst of enthusiasm following statehood, population growth in Iowa settled down into a pattern prescribed by FORE.

Biography of a ghost town. In 1910, the town of Le Roy, Iowa, had a hotel, two general stores, two gas stations, a café, a lumberyard, a stockyard, and a railroad. By 1946 the railroad was gone and things started to look bad. Then, as shown in Fig. 66, FORE took over. After the school and post office were closed, businesses departed, real estate values plummeted 90 percent, and the town coasted to a stop. The FORE projection is for an ultimate population of zero.

Le Roy was not your typical ghost town, left after the mine closed, because there was no mine. Small towns were squeezed by introduction of the school bus, as small schools were consolidated into bigger ones, and it is not easy for a local mom-and-pop store to compete with a Wal-Mart in the city

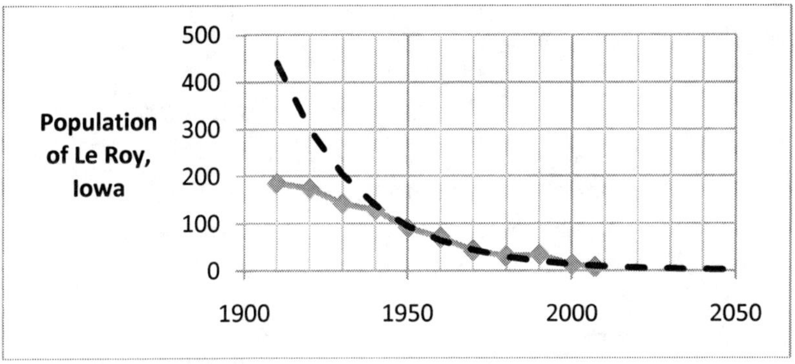

Fig. 66. The final fade-out of Le Roy went according to FORE except for an elderly lady who keeps cats.

And now Detroit? Between 1910 and 1950 Detroit was Boom Town, USA, owing to the rapid growth of the automobile industry. Then, as shown in Fig. 67, disaster struck, starting with entry of the Volkswagen. U.S. gas prices went up, and foreign car manufacturers had an advantage because they already were used to it. Meanwhile Detroit stayed focused on big cars with tail fins and heavy breathing. Labor unions demanded large benefits that it was assumed would go on forever so Detroit stripped its gears. It borrowed to try and keep up and then took bankruptcy. The population plunged two-thirds until about all that was left was a high crime rate and a dream of recovery.

Is Detroit only Le Roy on steroids? Towns that fade away usually don't come back, but cities do better.

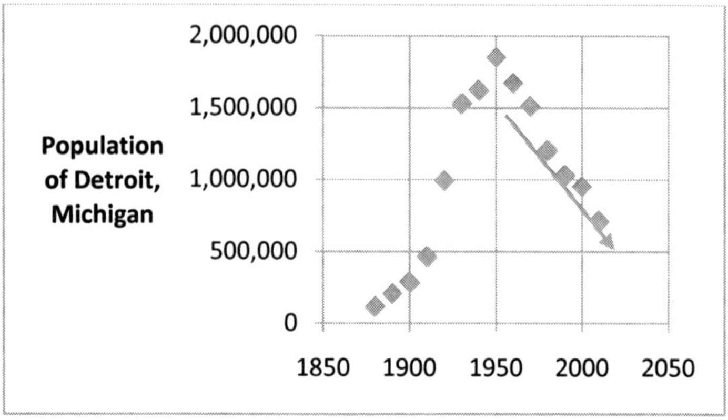

Fig. 67. So far the decline of Detroit has not followed FORE but maybe that's good news. Sometimes when something bad happens something good happens.

Pass the pizza

In 1922 the *New York Times* reported that the population of Italy was growing faster than the population of the U.S., and the article was out-of-date as soon as it was off the newsstands. Fig. 71 shows that the population of Italy is leveling off according to FORE regardless of guidance from the Vatican. The FORE projection for Italy is for an eventual 20 percent increase to around 69 million. However, the most recent census shows a population decline. That also is happening in other European countries so tax collectors are getting worried. In Russia, both the birth rate and life expectancy are in a decline but it has not affected the enthusiasm for vodka. With regard to population growth Europe may be more mature than America, but it also should pay attention to immigration.

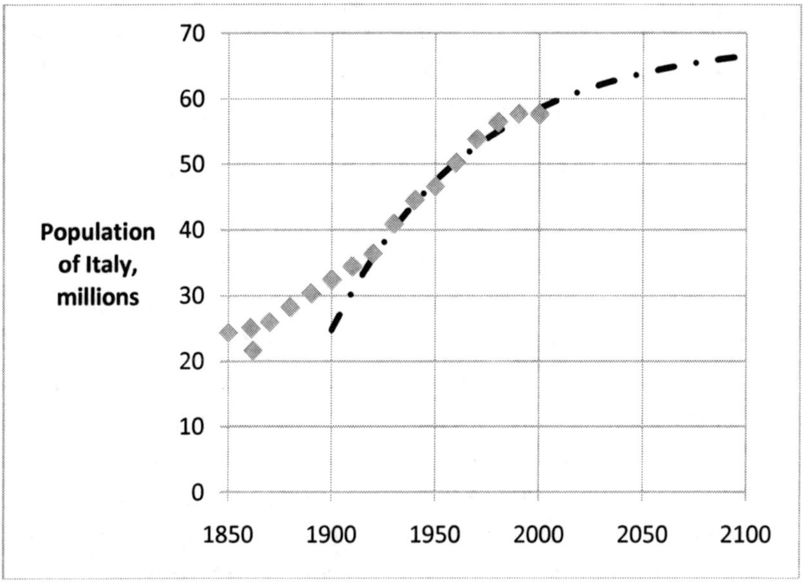

Fig. 68. Population growth in Italy has had some ups and downs. The dashed line shows a projection by FORE.

Peking duck

China used a heavy hand to stem population growth at about the same time that it imported capitalism and made stunning advances in manufacturing and growth of a middle class. At one time it was customary for a Chinese family to have six or eight children, and then the allotment suddenly was reduced to one with penalties for non-compliance. Now young people in China are beginning to like the idea, to the extent that someday there may not be enough young people to take care of the old people. It is easy for over- management to go too far.

India tried using a different kind of persuasion to control population growth, and it did not work. A young man who got himself fixed got a free transistor radio, so if you saw a young man standing around with a transistor radio in his ear, it probably was because he couldn't sit down. A problem with that is a dead battery can get a boost using jumper cables.

On becoming extinct

Dinosaurs, woolly mammoths, dodo birds, passenger pigeons, Tasmanian tigers, and early species of man all became extinct. One thing about extinction: No species can make the same mistake twice. For many thousands of years Neanderthal was on the top branch of the family tree and then the branch broke off. Some Neanderthal DNA did manage a lemur-like leap to another branch, but not enough to maintain a brow ridge. There are over 1000 species of jellyfish, and only one species of man. There are over 200 species of monkeys, and only one surviving species of man. Maybe what is needed is Viagra for the brain.

Fig. 69. One can't ignore the facts.

Chapter 9. Amazing Maize

The Aztecs had never seen a man on a horse so they were assumed to be inseparable, and the blend of sweaty odors just about clinched it.

Corn, wheat and rice are staples of humanity, along with beans, potatoes, oats, onions, bananas, peanuts, manioc, taro, noodles, turnips, radishes, nuts and couscous. Corn is by far the leading food crop. In fact, all that Biblical fuss about unleavened bread probably would have lost its punch if they had ever tasted some decent tortillas.

In 1519 Hernando Cortes and his fellow adventurers put on their tin hats and hubcaps and sallied forth across Mexico looking for gold. They also found corn and peanut butter, but they found gold manifestly more interesting. Cortes' translator was an Indian girl named *Malinche*, whom he tricked into fishing for the cod under his codpiece so he could show her a thing or two.

The Aztecs had never seen a man on a horse, so they were assumed to be inseparable, and the blend of sweaty odors just about clinched it. They assumed that Cortes was their No. 1 God named *Huitzilopochtli.* It is pronounced like it sounds.

Huitzilopochtli (pronounced like it sounds) was not just an ordinary god; he was the god who was in charge of winning wars and keeping the sun regular. The Aztecs therefore gave Cortes anything he wanted, and like any civilized individual, he picked gold. In return he gave them missionaries, smallpox, and opportunities to work in the mines if they didn't what their hands cut off. If Cortes had fallen off of his horse it probably would have changed the course of history.

Cortes sent the gold back to Spain and was anticipating an important promotion but all he got was a governorship and an urgent request for more gold. King Phillip II needed the gold to finance a mighty armada of ships so it could sally in the English Channel and teach the English a thing or two about Catholicism. In 1588 most of the ships sank during a storm, which the English interpreted as "God Saved the Protestants." That later was modified to "God Save the Queen" to try and ease Irish sensitivities. When that didn't work they sent in the troops.

The Aztecs pioneered heart transplant surgery but unfortunately never got past the first part. They prepared their patients by painting them blue, presumably so the color change during the operation would be less noticeable. The operations were performed by High Priests because they did their best work while high on top of a pyramid, and their success rate is what led to the expression to be "bumped off." It was heavy responsibility so whenever they ran short of captives they had to go get more.

Cortes was genuinely annoyed by this cruel and inhuman procedure, and he demanded that it be stopped because where he came from it was customary to burn people at the stake, which gave them the opportunity to repent and tell where they hid their money.

America the sweet and beautiful

While Europeans were dining on hardtack, oatmeal and sauerkraut, native Americans had sweet corn, potatoes, tomatoes, lima beans, avocadoes, chili peppers, chocolate, grapefruit, cranberries, blueberries, pecans and peanuts, plus meat, fish, crabs and oysters for extra flavor. They mastered the art of selective plant breeding by turning a wild grass into corn, so these were not primitive people. It has been claimed that they did not have the wheel but they did have the wheel, and used it for a calendar. The Mayan calendar slipped a cog when in 2012 people said it predicted the end of the world and some people were disappointed. Some were so convinced it would happen that they went ahead and checked out ahead of time, so for them it worked. It was a grave mistake.

The land of corn. Corn is corn flakes, popcorn, corn-on-the-cob, corn chips, corn curls, corn starch, corn syrup, corn oil, corn meal, corn bread, corndogs, corn tacos, corn pone, corned beef, corn liquor, polenta, dog food, hog food, and corny gags on TV. As shown in Fig. 70, the annual world production of corn is about five times that of wheat and a million times that of rice. Corn constitutes over 90 percent of the cash value of all grain crops grown in the U.S., and most of it comes from Iowa, northern Illinois and southern Minnesota. Two major exports from the U.S. are corn and money.

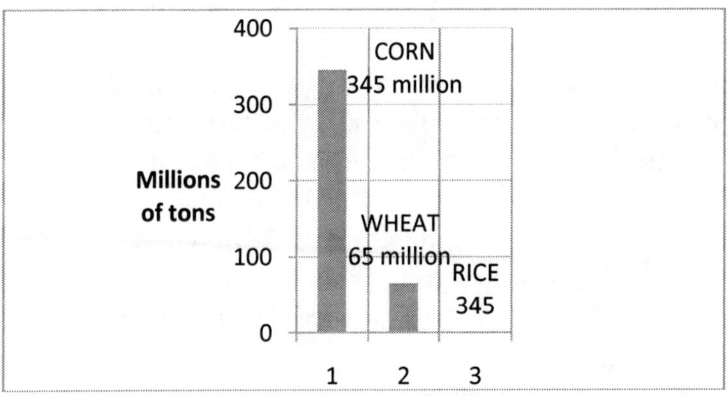

Fig. 70. World annual production of corn, wheat and rice. Note that the rice figure is not in millions. U.S. Dept. of Agriculture.

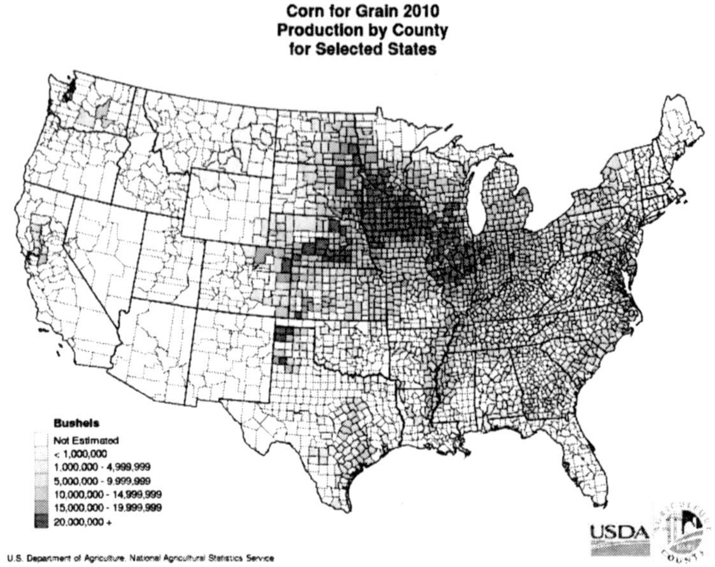

Fig. 71. Corn Belt, U.S.A. U. S. Dept. of Agriculture.

About half of the corn grown in the U.S. is recycled into pigs, cows and chickens. Another 20 percent is exported, and the other 30 percent goes into everything from plastics to soda pop.

Ethanol action. Fermentation is as natural as tooth decay, and as discussed in Ch. 6, corn liquor can make a snappy addition to gasoline, and is not a fossil fuel. One bushel of corn, 56 pounds, can make about 8.6 pounds (2.9 gallons) of ethanol, plus 16 pounds of animal feed. That is over a 40 percent utilization without counting cornstalks, leaves and cobs that with additional processing can make more alcohol.

An acre of corn currently can be processed into about 520 gallons of ethanol, enough to keep a car going for about year. Since there are many more cars than there are acres of corn, petroleum interests should not get too concerned. The percentage of alcohol in motor fuel is creeping upward, first starting at 10 percent and going to 15. With some adjustments in automobiles it can go as high as 85 percent, but at some sacrifice in miles per gallon.

Slash and burn

Early agriculture involved *slash and burn.* Today's equivalent is *bulldoze-and-burn*. Red-as-rust tropical soils can only tolerate a few years of crop production when it is time to move on and do more slash-and-burn. That can work if there is enough time to allow natural vegetation to come back and rejuvenate the soil. Otherwise, Sahara. At one time North Africa was the breadbasket of Rome. Now it is just North Africa.

Westward expansion. Westward expansion in the U.S. actually skipped over some of the most productive soil in the world, a blanket of soil that was laid down by continental glaciers. The tall grass that extended to the horizons put people off; they assumed that if a soil was any good it should grow trees. The prairie soils would have grown trees, but grass fires kept them burned off. Now the trees are filling in.

The Plains that broke the plow. Another problem with the black, organic prairie soils was that they were too wet and heavy to plow. Plows had been made out of wood ever since the dawn of agriculture so when settlers encountered wet, heavy soils they moved on to places where grass was shorter because the soils were drier, Then when fields turned to dust and blew away, people moved to California, which Will Rogers wryly observed probably increased the I.Q. in both places.

The plow that broke the Plains. In 1837 an Illinois blacksmith named John Deere invented the steel plow, which is not flat like a board but has a curve like a gambler's smile. Its lower knife-edge cuts down and pulls the plow down into the soil so it doesn't just scrape over the top, and the spoon shape lifts the soil and rolls it over sideways. That made the heavy, rich soils of the Midwest vulnerable, and soon over 10,000 of the new plows were being sold every year. It was the beginning of an industry that still remains firmly planted in the corn belt.

Crop rotation: the Arab connection. Crop rotation was used by Arabs for a thousand years before anybody else knew about it. Changing crops from year to year not only helps to rejuvenate the soil, it keeps the weeds and insects confused. In Europe crop rotation tried out wheat one year and oats the next, so choices were dark bead or oatmeal. England went for diversification with rotations that included wheat, turnips, barley, and clover. The U.S. went for corn, corn, oats and meadow. Beans, including soybeans, grab nitrogen out of the air and tuck it away in nodules in the soil to create a natural fertilizer. All that was necessary was to find some uses for the beans, after which a rotation went to corn, soybeans, clover and alfalfa with chicken every Sunday. Soybeans are used for cooking oil and to make everything from paint to linoleum.

Purveyor of peanuts. George Washington Carver, the son of slaves, got his degree and briefly taught at Iowa State College before going on to the Tuskegee Institute. He championed new uses for peanuts because, like beans, they put nitrogen into the soil.

Nitrogen with no waiting. Corn production has become so profitable that nitrogen now is pumped directly into soil as ammonia. Pure ammonia It is like household ammonia without the water, and has to be kept under pressure. The use of ammonia can lead to bumper crops but there is a bump in the bumper, because excess nitrogen ends up in lakes and rivers, where it causes algae blooms and pollutes drinking water supplies. Excessive amounts of algae in the Gulf of Mexico die, sink, and take out the oxygen, resulting in "dead zone" for fish and shrimp. This probably will continue as long as a primary objective of chemical companies is to keep selling more fertilizer.

An arranged marriage. A major contribution to corn production came out of research by Henry A. Wallace, who later became Vice President of the U.S. Wallace observed that corn plants are self-pollinating, which is a little like masturbation so prospects for change are limited. This can be prevented by simply pulling off the tassels, which are the male parts. That forces new match-ups called hybrids. The only drawback is that the hybrids are sterile, which is not unusual. A mule is an example; it is created when a horse gets romanced by some jackass and then comes to a full stop, nature having discovered her mistake. Hybrid corn is big business because a new batch of seed is required every year, by forcing selected corn plants into arranged marriages. One result is over a four-fold increase in productivity, Fig. 72.

In addition mechanical corn pickers allow rows to be planted closer together, and chemical fertilizers, pesticides and weed killers increase efficiency. Genetic modifications shortcut hybridization and show that all living things are related. Eventual caps may be the availability of water and intensity of energy from the sun, and researchers are working on the water problem.

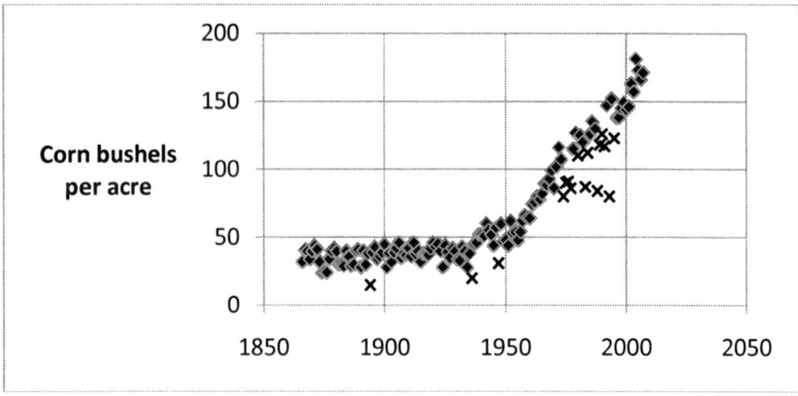

Fig. 72. Corn productivity in Iowa.

The movie monster. In 1818 a science-fiction novel called *Frankenstein* was written by 21-year old Mary Shelley. It featured an ill-conceived organ transplant, and the book and subsequent movie evoked a fear that is unlikely to ever go away. A similar fear is created from mingling genes in plants, which is a little like trying to unscramble eggs to make a better waffle. As populations increase it may be a key to survival.

Irrigation and all that follows. Irrigation puts water into the soil, and then, particularly in dry climates, the water

Fig. 73. Sometimes there is not enough room in the bins. Photo is by Tom Whitney and courtesy of Mark and Cheryl Roberts.

comes back up and evaporates. A difficulty is that the flow reversal brings up salts that concentrate in the soil. It is an acceleration of a natural process that makes "caliche," a surface layer of soil that is cemented like limestone and has the look of a desert. Irrigation is mainly used in places where there is a shortage of rain, so a second effect is lowering of the groundwater table, after which everybody registers surprise. Wells can go deeper but there are limits. Deep wells may drawing on water that has been there for many thousands of year and is not readily replaced.

Agricultural engineering

In 1793 Eli Whitney invented the "cotton gin," although it is unclear why he called it gin. It tears cotton apart so the seeds fall out, and it contributed to the industrial revolution.

In 1837 Cyrus McCormick invented a reaper that reaped the wheat so fast that it did the work of three men. In 1850 Edmond Quincy invented the mechanical corn picker, and then Hiram Moore came up with the idea for a "combine" that combines picking and harvesting. Now corn is picked, shelled and dumped directly into semitrailer trucks that haul it to a storage facility. The pickers have huge prongs that stick out and look like they could stab their way into a bank vault. The machines gobble up corn plants in at one end and spit out corn at the other. Agriculture is big business.

FORE and the future of corn

Fig. 74 shows a FORE treatment of corn productivity during good years in Iowa. The graph is like a fishing pole; any change in the direction of the handle is magnified at the tip. The projection is for an ultimate peak corn production of about 400 bushels per acre, which would more than double the current average. Coincidentally the current world record is 393.74 bushels of corn per acre, set in 1999 by the winner of the Iowa Master Corn Grower Contest.

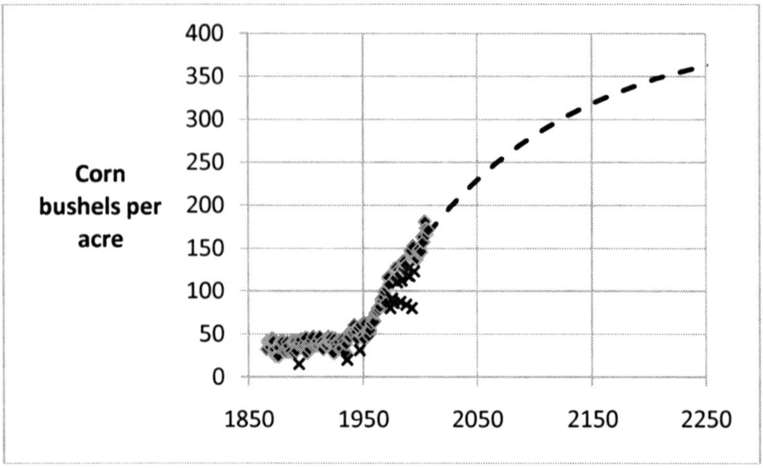

Fig. 74. A FORE projection of future corn productivity that may be a bit of a stretch but the ultimate value coincides with the current record production.

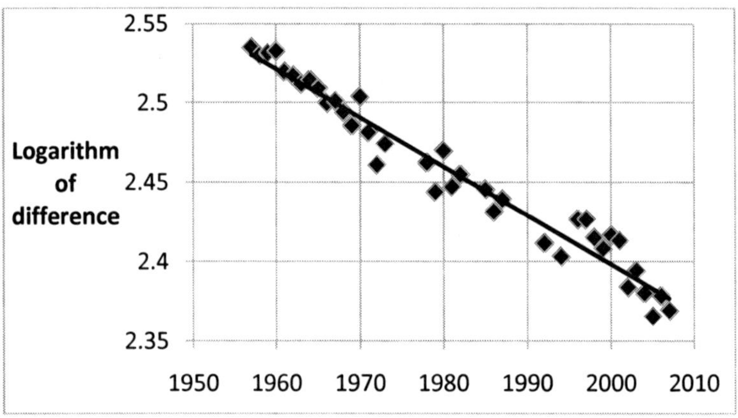

Fig. 75 The graph of FORE in the previous figure is straight but the R^2 is a bit low at 0.96.

The Greening of wheat

Wheat was created as a result of selective breeding about 4000 years ago in the "Fertile Crescent" of the Middle East. This now is acknowledged with a French roll called a *croissant,* or it may only have been invented by French baker with a cauliflower ear.

Wheat production was being decimated by a fungus called *wheat rust,* so an agricultural scientist named Norman Borlaug went to work on it. He used some of the principles that had worked for corn and applied them for Mexican wheat. It must have worked because as shown in Fig. 76, in 40 years, from 1960 and 2000, Mexican wheat production went up by a factor of four, which rivals the rate of increase in productivity of corn.

Wheat production also has almost doubled in India and in Pakistan, but population growth still is skating along a

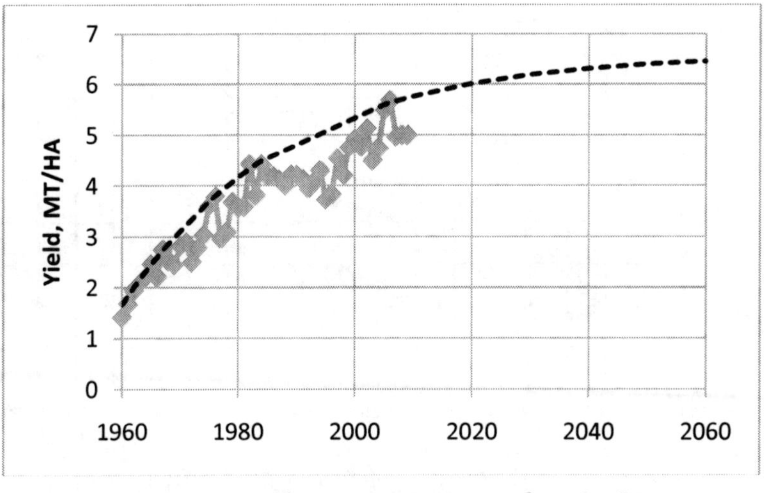

Fig. 76. Wheat production in Mexico after the "Green Revolution" when selective breeding created resistance to wheat rust. FORE is for the peak years.

ragged edge. Nevertheless the Green Revolution is credited with saving over a *billion* human lives, and Borlaug was awarded the Nobel Prize and helped to establish the World Food Prize administered by Iowa State University.

Wheat rust now is evolving and making a comeback, so the only way to keep up will be with continuing research. Even the Green Revolution can't be taken for granted.

The Greening of rice

The dramatic successes in the production of corn and wheat have led to similar efforts for rice. In 1960 the International Rice Research Institute was established in the Philippines with grants from the Ford and Rockefeller Foundations. As shown in Fig. 77, rice productivity already was high in Japan, but it still has increased over 50 percent.

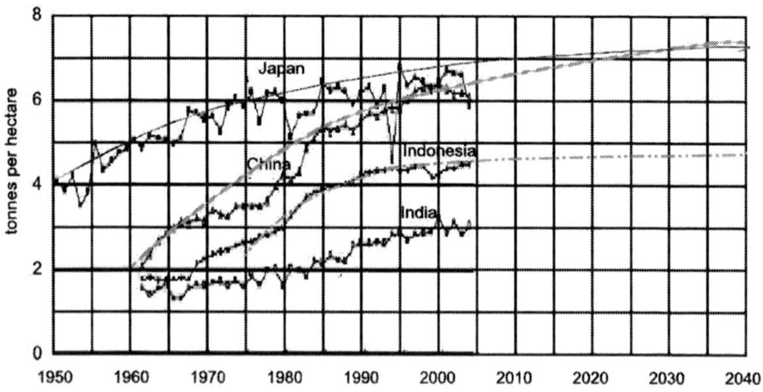

Fig. 77. A FORE treatment of peak rice production in Japan, China, and Indonesia. India lags. Data are from Takeshi Horie et al., 4th International Crop Science Congress, Brisbane, Australia, 2004.

In China, rice productivity increased over 300 percent after Mao Zedong died and scientists were allowed to do what they do best, instead of laboring in the fields. Chairman Mao apparently had this thing about laboring in the fields. The FORE projections for maximum productivity in China and Japan both come out the same, 7.5 metric tons per hectare or 3.3 tons per acre. Productivity in Indonesia and particularly in India have a way to go.

Buy the sea

Fish has always been good but it has become even better since it no longer has to be eaten on Fridays. There is a catch--the supply is limited.

Fig. 78 shows the codfish catch from an area off the coast of Labrador and Newfoundland. In the 1920's

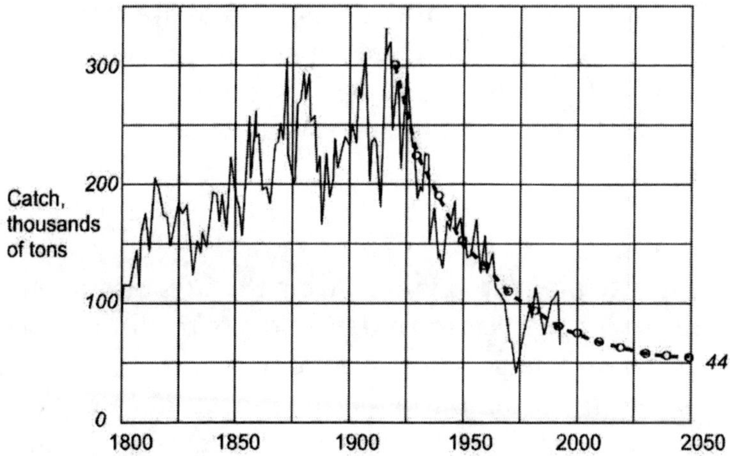

Fig. 78. The FORE projection is for an 85 percent decrease in the codfish catch in one area as a result of overfishing. Data are from Lear and Parsons (1993).

fishing was from small boats with hooks and lines. It yielded some fish that weighed over 100 pounds. Then fishing became industrialized as trawlers came from distant places to drag heavy nets across the sea bottom. That not only netted the fish, it ruined much of the habitat. It was a classic case for FORE, with a predicted ultimate production dropping from over 300 down to 44 thousand tons per year. After 1990 steps were taken for a recovery, but so far the success has been limited.

Lobster action. At one time lobsters were so abundant and so ugly that they were used for fertilizer. They were so cheap that hired help insisted that they not be served more than a couple times a week. Now that has changed. Lobsters now are so pricey that little chunks are dunked in melted butter to make them go farther. Every Eden has its limitations.

The good news is that the lobster traps are starting to fill up again, but the price of course remains elitist. Lobster catches now are having some record years, possibly attributed to climate change. It's a price we may have to pay.

Hard times

Contrary to popular opinion during the Great Depression of the 1930s there was too much food, not too little. Food supplies were up so prices were down, so farmers could not afford to stay in business, and more money could be made on the stock market. As Will Rogers pointed out, in order to make money on the stock market one should buy stock, wait until the price goes up and then sell it, or if the price doesn't go up, don't buy it.

Chapter 10. Somewhere west of Gibraltar

"The first thing we know we won't know a thing."
—Socrates, on the Euro.

The Greek philosopher Socrates created a unique way of teaching. He would ask his students questions and let them figure out the answers. That gave Socrates the reputation of being very smart, because he got all of the answers with only a whisper of effort. All he had to do was listen and drop a little praise here and there to keep things moving. Now it is called assigning a thesis topic.

Socrates' most famous student was Plato, who had so much respect for his mentor that he decided to write down everything that Socrates said. That included some zingers like "Take my wife—please," which shows how old that one is. Socrates advised, "By all means marry. If you get a good wife you will be happy, and if you get a bad one you will become a philosopher." Plato became a philosopher.

Socrates is said to have made fun of politicians when all he did was report the facts. Then he made fun of the Greek gods where there was a shortage of facts so he was directed to drink hemlock, which was a kind of gin with an attitude. Socrates drank it and died but he had the last laugh, because he knew that his critics had been so vocal about it that they could not possibly get re-elected.

Plato thought that he remembered Socrates mumbling something about a "lost continent" but he actually may have been was looking for a lost condiment, namely some Tzatziki sauce for his gyros. Plato probably thought that

that "continent" made a better story, and people have been looking for a "lost continent" ever since.

Plato called the continent Atlantis, and indicated that it sank into the sea somewhere west of Gibraltar. He also reported that it had hot-and-cold running water so two possibilities that immediately come to mind are Iceland or Yellowstone Park. Old Faithful was a lot younger then and it may not have been so faithful.

Iceland? Iceland is a volcanic island that has hot and cold running water but it is not really lost, only just a little out-of-the-way. The hot water is used for geothermal heating and the cold water comes from glaciers. If Iceland did sink it must have come back up, in which case there should be some evidence like old coral reefs and shipwrecks like Noah's Ark.

Iceland is on the *Mid-Atlantic Ridge,* a steamy seam where the bottom of the Atlantic is coming apart. It runs roughly north-and-south and plays a prominent role in a concept called "continental drift."

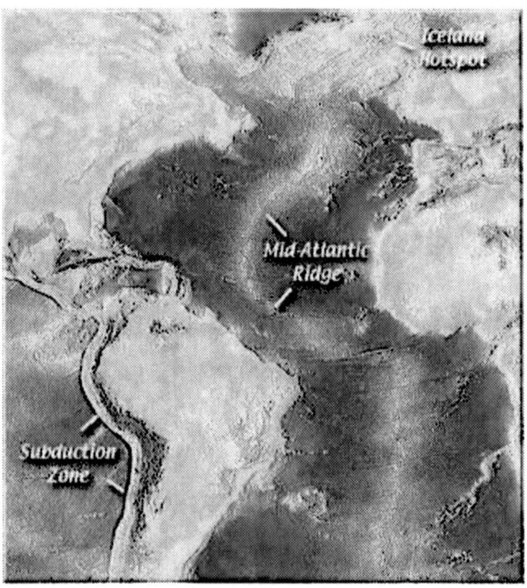

Fig. 79. Take away the oceans and it looks like the continents split. the line of separation is a line or volcanism called the Mid-Atlantic Ridge.

Did the continents really drift?

In 1912 a German meteorologist named Alfred Wegener was sipping his schnapps while looking at a map such as in Fig 79, and observed that the eastern edge of the South America is almost a perfect fit with the western edge of Africa, with the hump of Brazil on the one side being tucked up under the armpit of Nigeria on the other. He then discovered that rocks on both sides of the ocean are identical, and that fossils are the same up to a certain point in time when they diverge into different species.

Wegener therefore thought it reasonable that at one time the continents were joined together and then came apart like pieces in a jigsaw puzzle, and in 1915 he wrote a book about it. His idea was immediately pummeled by virtually all of the leading geologists of the day, but that may have been because Wegener was a meteorologist and the geologists welcomed a change from pummeling each other. They even held a special conference so they could take turns confirming that Wegener was an idiot. It did not bother Wegener because of his prior experience as a meteorologist.

Wegener was so cool that in 1930 led an expedition to Greenland but things did not go well for him, and he died with his skis on. He probably had a heart attack, but since he was a heavy smoker he may only have been dying for a smoke.

Wegener called his concept "continental drift." At best it was regarded as anecdotal and a little mixed up, like Snow White and the Three Little Pigs. Then during the 1950s new data started filtering in that supported his concept. At first it had to do with reversals of magnetism in rocks in the sea floor. A clincher was the discovery of "seamounts," submerged volcanoes scattered over the floor of the Atlantic that are progressively older the farther away they are from the Mid-Atlantic zipper. The bottom of Atlantic Ocean therefore is like two conveyor belts running away from the Ridge. Despite all of the professional jealousies and discourtesies, it turned out that Wegener was right.

Continental drift now is called *plate tectonics* so it can include discrete pieces of ocean bottom that also are moving around and bumping into things. When a continental humps an ocean plate, the result is a mountain range on one side and an ocean deep on the other. Two continental plates make high mountains, the Himalayas. Plate boundaries rim the Pacific Ocean to create a "Rim of Fire" marked by volcanoes. The Hawaiian Islands were and are being created as a plate drifts over a "hot spot" that burns through.

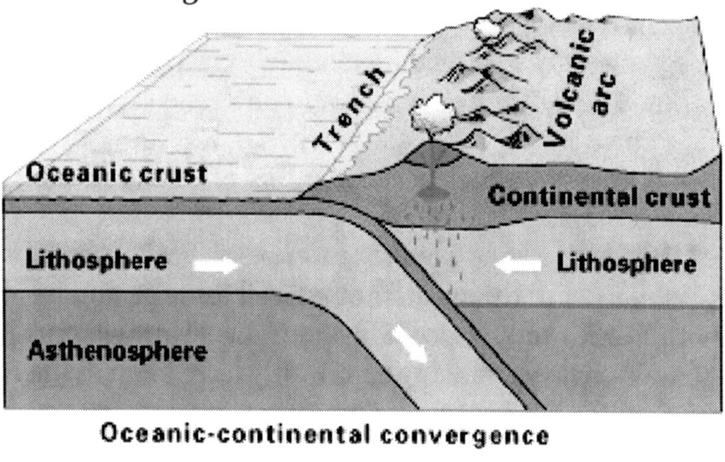

Fig. 80. California style crunch. Movement is not steady but involves "stick-slip" that creates earthquakes. U.S.G.S.

GPS measurements show that movement rates vary from 2.5 to 15 cm (1 to 6 inches) per year. Much of the denial by early geologists was based on a lack of any mechanism that might do this. Now geophysical measurements indicate that the continents are being transported by very slow heat-induced convection currents in pasty underlying rocks.

Can continental drift and plate tectonics account for Atlantis? It might, but mostly the land just comes up.

Ancient cities

Alexandria is a city in Egypt, and part of Alexandria really did sink. The city was founded by Alexander the Great and was a favorite of Cleopatra. She probably was curious about what made Alexander so great. The city in part was built on soft soils of the Nile Delta, which was not good. Alexander died in Iraq at the age of 32. There should be a lesson in that.

Might Alexandria be Atlantis? Alexandria was founded in 331 BC so Plato could not have known about it because he already was dead.

Limantepe. An ancient city was recently discovered underwater off of the coast of Turkey. It either had to sink or the water level came up. It dates from around 700 BC, about 400 years before Plato, which would give plenty of time for the story to be kicked around and exaggerated. It was not west of Gibraltar, but since the earth rotates west to east maybe something slipped.

Troy. Troy was a bustling city prior to the incident with the wooden horse. Troy was described by Homer but was assumed to be a work of fiction until 1868 when its ruins were discovered by a German archeologist named Heinrich Schliemann. Might Troy qualify as Alexandria? Not likely. Troy was on hard rock and could not sink.

Why cities sink

Cities usually are located next to water—along a river or close to the sea or both, which is asking for it. River floodplains occasionally flood, which appears to be a valid reason to call them floodplains even though that tends to be overlooked by commercial developers.

Floodplain soils often have intersecting loops of abandoned river channels or "oxbows" that are filled with soft clay. Inside of the loops is sand so it is important to distinguish. Sinking is so common in some areas the front steps go down instead of up.

Deltas occur where rivers dump their sediment load in quiet waters of lakes and seas so soil grains settle out like snowflakes on a quiet night. As more soil is deposited and pressure increases, water squeezes out and allows settlement.

Alexandria, Limantepe, Venice, New Orleans, and many other coastal cities are sinking. Modern buildings are supported on piles that extend down to bedrock so as the city goes down it looks like the buildings are coming up. Sidewalks tilt away from the buildings so pedestrians have to walk on a tilt or develop one leg that is longer than the other.

The science of sinking.

In the 1920s an Austrian engineer-turned geologist named Karl Terzaghi became intrigued with the mechanics of sinking, so he built a machine to try and make some sense out of it. His machine compressed samples of soil and showed that rate of sinking is controlled by time required to squeeze water out. The finer-grained and the thicker the clay layer, the longer it takes for water to be squeezed out. He took a hint from the dynamics of heat flow and wrote the basic equation. That's how analysis sometimes operates, and FORE is no exception.

A delta should be a good place to look for Atlantis, except that anything left laying around on a delta is likely to be covered up.

Sinking sand. Sandy soils can look safe but they also can have their quirks. If given a good shake, as by an earthquake, a loose sand will collapse, and if it is full of water it will turn into *quicksand!* The formal name is "liquefaction," because it turns into a liquid that is dense enough to float a person but not dense enough to float a building. Anybody caught in quicksand should relax and try to swim out. If that doesn't work, somebody should throw them a rope.

Might Atlantis have been sited on loose, saturated sand that was susceptible to an earthquake? It is possible, but there would require that it be built on loose, saturated sand and that there be an earthquake.

Fig. 81. Reelfoot Lake is on the Mississippi River floodplain in western Tennessee, and suddenly sank after the 1812-13 New Madrid earthquakes. Author photo.

Sinking related to underground caverns. Limestone caverns form where rainwater filters down through soil and concentrates at the groundwater table, where the dissolved carbon dioxide creates and acid that dissolves away the limestone. As river valleys erode deeper so the groundwater table is lowered, it can create a new level of caverns. Vertical connecting channels are properly referred to by spelunkers as "glory holes."

Might Atlantis have been over a cavern? It presumably would have had to have been a very large cavern or it would only take out a chunk of suburbs.

Mine collapses also can swallow sizable bits of terrain, and it is shallow mines that are the greatest risk. That is because deep mines are filled with loose rock when the roofs cave in.

Sinking sideways

Landslides induce lateral when they downhill. They can be triggered by prolonged periods of rain and by earthquakes. The main requirements are a hill and

Fig. 82. Curved shoreline where a chunk of Seward, Alaska, slipped off into the sea during the 1964 earthquake. USGS Menlo Park photo.

slippery soil. Might Atlantis have been on a landslide? There should be clues, and a convenient place would be a high bank next to the sea.

Sinking and FORE

In the 1990's an artificial island was built out into the sea to support a major airport in Japan. It was expected to sink, and it did, so FORE took a shot at it. A typical result is in Fig 83, which indicates an eventual total amount of sinking of 16.4 m (54 ft) at this particular location. That is about equal to the height of a 5-story building.

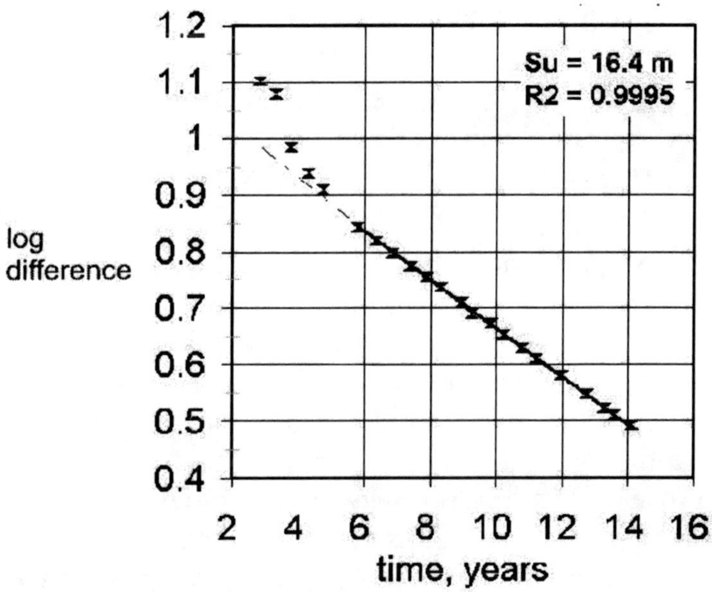

Fig. 83. FORE applied to settlement of the Kansai International Airport, Japan. The high R^2 value testifies to accuracy of the data. The data are courtesy of Prof. Koichi Akai, Univ. of Kyota, Japan, and the Kansai International Airport Commission. The reference is in the Reference List, Handy (2003).

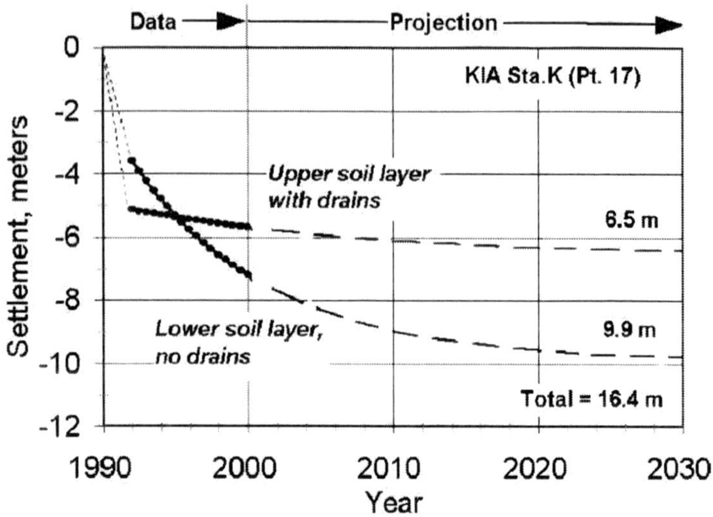

Fig. 84. FORE and sinking of the airport from compression of two different soil layers. The upper layer was penetrated with vertical columns of sand to encourage faster drainage.

What to do now; just let it sink? The ratio of fill thickness to sinking is almost exactly 2-to-1, so two feet of fill will be needed to compensate for every foot of settlement. The Kansaii International Airport is the second most important in Japan, and the project is being expanded.

An elderly professor severely criticized this analysis after he got the logarithms messed up, and claimed that it was only "a matter of semantics." Mark Twain pointed out that semantics are what distinguish lightning from a lightning bug.

Might Atlantis have blown up?

Maybe Atlantis did not sink; maybe it just blew up and disappeared. That happened at Yellowstone about 640,000 years ago, and the area still is simmering. A huge explosion occurred in 1600 BC at Santorini, an island in the Aegean Sea south of Greece, and Socrates probably knew about it. All that was left after the explosion now is a rim of jagged islands around a crater filled with cruise ships.

Might Santorini be Atlantis? A problem is it did not sink; it blew up. A similar explosion in 1883 took the island of Krakatoa in Indonesia off the map. Now the volcano is erupting and the island coming back. Floating debris from when Santorini blew up indicates that the tsunami was as 100 m (300 ft) high. It washed away the Minoan civilization and it never recovered.

Fig. 85. Santorini, one side of where the mountain used to be. The white layer on the surrounding hill is volcanic ash. It was a mighty explosion that could have created a condition for Atlantis almost anywhere around the Mediterranean. Author photo.

Self-inflicted sinking

Cities require fresh drinking water, and the easiest way to get it is from wells. Romans brought water in from nearby hills with their famous viaducts, which eventually led to the downfall. All that was required was to cut a chunk out of a viaduct when nobody was looking and shut off the water.

Water taken out of the ground also can create some disadvantages if it is not replenished by rainfall. A main disadvantage is soft ground will sink. Mexico City and Venice are examples.

As a groundwater level goes down it's like pulling the plug in a bathtub: contents get heavier, and when they get heavier they are likely to sink. Might that explain Atlantis? Probably not; they didn't have well pumps.

Recovery. What goes down should come up, so why not just pump water back in to replace what was pumped out? That sounds like such a good idea that it was tried, for example in Venice, and it did not work. The problem is that soil particles that are pushed together stay together. They develop a genuine friendship for each other and there is no easy way to tear them apart. Lifting can be achieved by pumping in a cement grout but that is many times more expensive than plain water.

Multiple working hypotheses; where to look for Atlantis

The above scenario introduces an important scientific procedure called the "Method of Multiple Working Hypotheses." Instead of pinning all hopes on a single explanation, equal consideration is given to all possible explanations and some that are not so possible. That is done to eliminate bias, as there is no more enduring than a pet theory. The process was formulated by an American geologist, T. C. Chamberin. Here is how it works:

• Maybe Atlantis was built on a delta where the soil was only strong enough to support its own weight so when a city was built on top it sank. Look for a suitable delta.

• Maybe Atlantis was built on loose sand in a delta or floodplain and got jarred by a major earthquake. Santorini could have provided the earthquake.

• Maybe Atlantis was built on an area that slid off into the sea, possibly prompted by an earthquake. In that case Atlantis should still be there but a little wet. The shoreline should have an arc-like shape as shown in a photo at the end of this chapter.

• Maybe Atlantis sank for some other reason, or it might not have sunk at all. There might not even have been an Atlantis. It could only be a matter of literary license. Happens in politics all of the time.

Fig. 86. Venice. Humorous writer Robert Benchley wired his editor, "Streets full of water; please advise" Photo by E. G. Ferguson.

Chapter 11. Horsepower

It was when the horse-and-buggy became a popular mode of transportation that it became fashionable to wear a bustle as everybody was accustomed to looking at the back end of a horse.

It is impossible to discuss horsepower without giving some consideration to the power of a horse. The first horse was a compact American model called *Eohippus.* It was not very big and had to run fast in order to survive. Early models kept stubbing their toes so the horse gradually evolved into one that now runs on its big toenails.

Man came to America cold and hungry from Siberia and discovered a happy hunting ground. He then happy-hunted many large animals to extinction, animals like the wooly mammoth and giant sloth that had assumed they were too big to fail.

Elsewhere horses managed to cut a deal: They would allow man to ride and guide, but only if he would concentrate his appetite on other options. That led to invention of the chariot and of the wheel. It was when the horse-and-buggy became the popular mode of transportation that it became fashionable to wear a bustle, as people were so accustomed to seeing the back end of a horse.

Horses returned to the Americas with the Spanish *Conquistadors,* and quickly were adopted by native Americans to hunt buffalo so it no longer was necessary to try and chase them off a cliff. A man's life came to depend on his horse, so there was only one way to rehabilitate a horse thief.

Horses went through a series of arranged marriages to develop specialty models such as draft horses, race horses, quarter horses, rocking horses, saw horses, horseradish, and hors d'oeuvres, to mention a few. Horse sense says don't play the horses.

When knights returned from the Holy Land they brought back Arabian horses and horse racing. By the 1700s horse racing was big in England. The stopwatch was invented in 1776, and timing of races began in 1848.

The Epsom Derby

This was the father of all of the Derby's. Winning times through the years are shown in Fig. 89, with FORE applied only to the record times. A sharp dip of about 5 percent occurred in 1900. It has the appearance of a *major event,* so what was the event?

Straight in the saddle. Paintings and lithographs from the 1800s show riders sitting stiff and upright in the saddle, whereas now the jockeys ride hunched over like a monkey riding a banana. The forward-looking posture was introduced to Britain by rash Americans who had no feeling for tradition, and as further aggravation they kept winning races.

Fig. 87. Horse race c. 1866 according to Currier and Ives, with riders sitting upright on the saddles.

Now jockeys don't just sit and go along for the ride, but lean forward and stand in the stirrups. That gives better streamlining but more importantly, the jockey becomes part of the running machine, a superb athlete in one of most dangerous of professions because if things go wrong, the consequences can be smashing.

Fig. 88. Leaning forward and carrying weight on the stirrups increases speed by allowing a jockey to synchronize his movements with those of the horse.

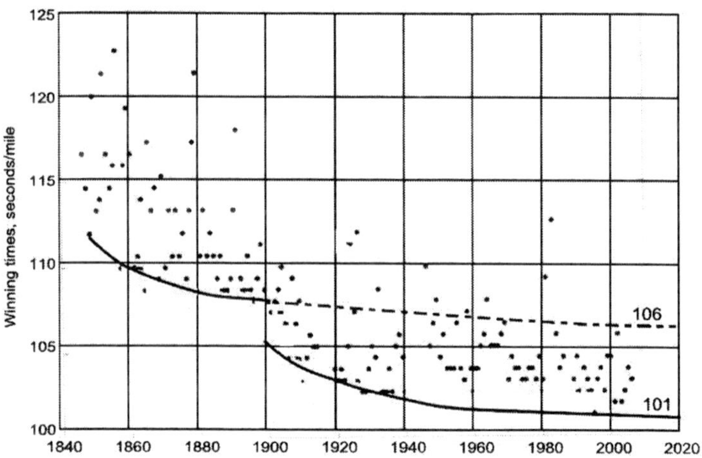

Fig. 89. A FORE treatment of fastest times at Epsom attribute a 5 percent increase in speed to work by the jockey.

Making the turns

Going around a turn involves some serious jockeying because the inside distance is shorter, and as suggested by the inclinations in Fig. 90, going around a turn is like trying to climb a hill sideways. Horses have to lean into the turn so centrifugal force adds to their weight. Cars going around a curve experience the same force so more weight goes on outside wheels so cars tilt the wrong way and tend to skid. A horse also can dig in its toenails.

Fig. 90. Horses instinctively lean into a turn to counteract centrifugal force and keep from tipping over. That also increases weight on the hoofs. Horse Racing Tips, http://www.toplineperformancehorses.com/

On the fast track

The fastest time for the Epsom Derby is 2 minutes 31.33 seconds, set by Workforce in 2010. That is an average speed of 35.8 miles per hour (57.6 km/h) for a distance of 1.505 miles.

The fastest time for the Kentucky Derby is 1 minute 59.4 seconds, set by Secretariat in 1973. That is an average speed of 37.7 miles per hour (60.7 km/h) 1973 for a distance of 1.25 miles, which is about 5 percent faster than the record at Epsom.

The Kentucky Derby distance was the same as for the Epsom until it was reduced in 1896. However, according to Fig. 91 the shorter distance did not result in a sudden change in the average speed, so any fatigue factor must be compensated by the time required to get out of the starting gates.

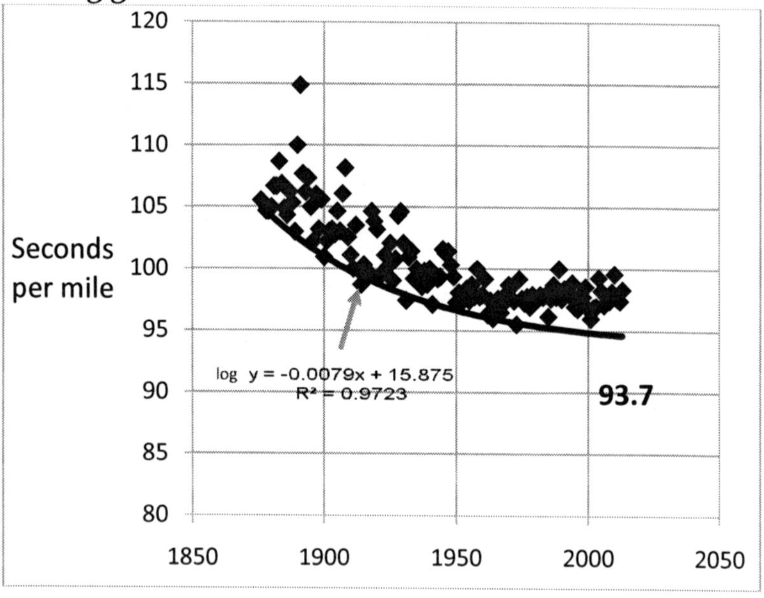

Fig. 91. FORE applied to record-breaking times in the Kentucky Derby indicates that shortening the track by a quarter of a mile in 1896 did not affect average speeds .

Some really fast horses

Only two horses have cracked the 2 minute mark in the Kentucky Derby. They are Northern Dancer ridden by Bill

Hartack in 1964, and Secretariat ridden by Ron Turcotte in 1973. In the latter race Secretariat was running in fifth place and then must have taken the insult personally. After he got cranked up he not only won, he set a new track record of 1 minute 59.4 seconds that still after 40 years it has never been equaled.

Secretariat also won the Preakness the same year, at 1 minute 54.4 seconds, and the Belmont, which is 1.5 miles, at 2 minutes 24 seconds. That gave him the fabled triple crown, and all three records still stand. At the Belmont, Secretariat won by an astounding 31 lengths, 248 ft (76 m). It was the fastest 1.5 miles ever officially timed for a horse running on a dirt track.

Secretariat was a super horse, but was he ahead of his time? Actually he was not because other records such as that of Northern Dancer come close, and Secretariat's time fits into the FORE analysis. Two horses that were ahead of their times were George Smith in 1916 and Twenty Grand in 1931, so let's also hear it for the old timers.

What about Seabiscuit? Seabiscuit was one of the most celebrated race horses in the U.S. He was small, had stubby legs, knobby knees and a difficult adolescence. He also did not like show-offs. However, his best years were 1937-38 when he was 5 years old, and the Derby and other triple crown races are limited to 3-year olds. It clearly is a matter of age discrimination. Maybe there is a concern that older horses may not make it.

Seabiscuit's most famous race was in 1938, a one-on-one against the highly acclaimed War Admiral. A year earlier War Admiral had won the triple crown. The first surprise was when the two horses ran neck-and-neck for most of the race. The second surprise was when Seabiscuit pulled ahead and won by four lengths. It was a proper subject for a movie that was nominated for an Academy Award even though the main love interest was the horse.

Which is faster, man or horse?

One of the most vexing questions of the 19th Century, aside from whether children should be allowed to work in the mines and women should be given the vote, was whether a man could outrun a horse. A racehorse can run the mile in about one and one-half minutes whereas a man takes four or more minutes so there should be no contest and the horse should win, but don't bet on it.

In the 1930s, the American runner Jesse Owens put it to the test in exhibition runs where the distance was 100 meters, and Owens always won. His only requirements were that the horse had to be a thoroughbred and have good hearing. When the starting gun went off close to its ear, the horse would jump back while Owens was off and running. A fortune teller who bet on the race was one happy medium.

The Wales connection. In 1980 the matter came up for some serious discussion in the Neuadd Arms pub in Lanwrtyd Wells, Wales, where it took a couple of pints just to be able to say the name. The race went 22 miles (35.4 km), probably in the hope that the horse might get lost but it didn't, and it won.

Fig. 92. Not everybody approves. bbc.uk.

The race started at the pub and ended at the pub so it turned out to be very popular and was repeated the following year. A horse still won. The race became a celebrated annual event, probably with more emphasis on the celebration than the event, and horses kept winning. Finally in 1989, a man won the race. He was riding bicycle. The rules probably were written in Welch.

The first human to win the man-against-horse race fair-and-square was named Huw Lobb, and he won in 2004. There were 500 runners and 40 horses, which probably created some confusion. Some contestants may even have run backward to try and confuse the horses so they were thinking they were running the wrong way. The prize was over $30,000 after a modest deduction to help pay for the celebration. A man won again in 2007, as nearly as anybody can remember.

Fig. 93. Next time no more stopping off to graze.

Chapter 12. How now, global warming?

Without global warming, places like Chicago still would be looking up at a mile of glacial ice. People who live there are reminded every winter.

So here is the deal, no argument: Heat causes water to evaporate, which is the principle of the clothes drier. If the drier doesn't heat it doesn't dry, so you call somebody. Clothes no longer are hung out on clothes lines because "hanging out" has taken on an entirely different meaning and involves a different set of qualifications. Now we are ready to talk weather.

Weather is hottest near the equator because it gets a direct hit from the sun, so air near the equator expands and rises. That is the principle behind the hot air balloon. A couple hundred years ago that was cutting edge so it shows how far we have come.

When a gas expands, it cools. Warm air rising therefore cools and loses part of its capacity to retain water so it makes clouds and rain, which is why the tropics need lots of umbrellas. However, when water vapor condenses it releases heat so the big, white, tall fluffy clouds that do a lot of heavy breathing. Sometimes updrafts keep ice particles suspended so they recycle to make hail. Airplanes go over or around.

Since warm air is less dense than cold air, it exerts less pressure and sometimes will collect itself to form a "low pressure area." When a low pressure area collides with cold air the relationship will be stormy.

Hurricane action. A low pressure area that drifts over a warm ocean picks up more heat and may graduate into a full-blown hurricane. During summers low pressure areas can be launched westward out of Sahara like they are being thrown by a pitching machine, then take in more energy as they cross warm waters of the Atlantic near the equator. Depending on the degree of success they can become hurricanes.

Why hurricanes spin. In the northern hemisphere hurricanes rotate counterclockwise because that is how they are supposed to go. Cars drive on the right side of the road also set up counterclockwise rotation so driving on the left is going against nature. Even rivers prefer to drive on the right in the northern hemisphere, as floodplains tend to be wider on the left.

The reason is coupled with rotation of the earth: Throw a ball and it goes straight, but by the time it gets to where it us going, the scenery has shifted, and at the equator the scenery is shifting at about a thousand miles per hour. The ball therefore appears to be deflected to the right. This is the *Coriolis effect*. It looks like a force and it acts like a force but it is not a real force; it is only an effect.

Since the earth is round, all scenery is not moving at the same speed. It goes by slower with increasing distance from the equator. At the poles all it does is stand still and look around. The imbalance can put a spin on a mass of air so it rotates counterclockwise in the northern hemisphere and clockwise over the southern half.

Coriolis also works for tornadoes, but since they are over land they don't keep picking up energy. About the best that they can do is create a few miles of utter destruction before they spin out.

Air going in circles also employs a trick used by ice skaters, as they pull in their arms and legs in order to spin faster. In baseball when a batter swings and misses he can say, "Oh, sh-t; must be Coriolis effect!" He's wrong. The best it can do is a tiny fraction of a millimeter.

Another "force" that is not a real force is gravity. Einstein reasoned that if a ball is inside of a box and both are falling, the ball will not feel gravity, so he proceeded to develop a space-time relationship. He had an inquiring mind.

Another way to play the game

In Spanish "*los niños*" means "the children." They play in the Pacific Ocean along the equator, where water temperatures oscillate up and down like a ball on a spring.

Oscillations are in the nature of things. In Fig. 94 the peak-to-peak change is about 4 C (7 F). Measurements made prior to 1950 were a bit casual because global warming had not been invented yet.

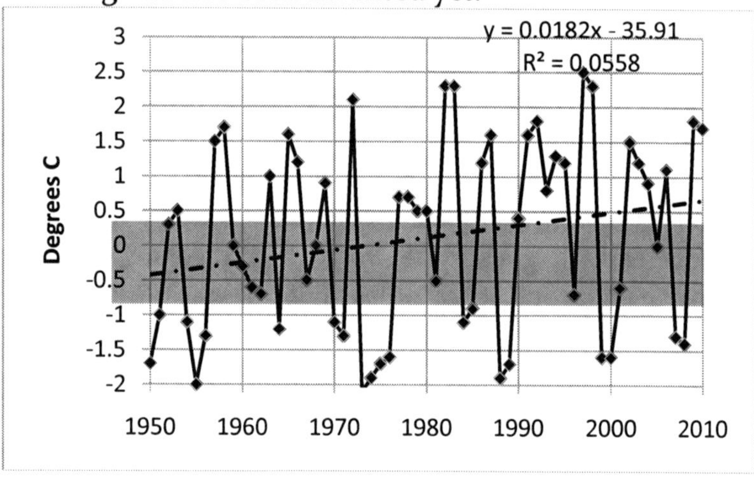

Fig. 94. Ocean temperatures in the Pacific along the equator. Temperatures outside of the shaded arbitrarily identify "los niños," hot for boys and cool for girls, which figures. NOAA, National Weather Service.

The *niños* are so abrupt that FORE has very little to say about it. However, a warming trend is apparent in this particular part of the ocean, increasing the average temperature about 1.8 C (3.2 F) in 100 years.

A better name for *El Niño* would be "*El Toro*" because of the way it bulls its way into South America, dumping on so much warm water that it ruins fishing. It was first noticed at Christmas. "*El Niño*" probably was a way to say "Holy Jesus!" without swearing.

After engage the ankles of the Andes they deflect and carry a message of heat and drought as far north as Canada. They may give the U.S. Southwest a bit more rain, but that is because there is so little to start with.

La Niña is the reverse of *El Niño* and is cool, with winds blow the other way so the fish come back. At the other end of the path they can set up a "Walker circulation" that dumps rain on Australia and Indonesia.

The SOI. The "Southern Oscillation Index" is a measure of the *Niños.* It is a corruption of "Sorry." It is defined from air pressures measured in Tahiti and in Darwin, and provides a good excuse to go to Tahiti and swing in a hammock.

Hot spots. Fig. 95 shows hot spots and cold spots identified from satellite and remote sensing measurements over land and over sea. Hot spots dominate, and cold spots are mainly close to Antarctica where huge chunks of shelf ice are breaking off, floating around and melting.

Arctic rainbows. Warming automatically becomes emphasized in Arctic because of the discrete temperature at which ice melts. In moderate climates moderate swings in temperature may not even be noticed as long as the air conditioner works, but in the Far North it can amount to sink or swim: The land sinks and the polar bears swim. A substantial part of permafrost, or what used to be permanently frozen ground, is ice, which means that when the ice melts the land sinks and ends up as a lake. Pack ice is melting off of polar seas, and already the bears are turning white with worry. Some may even be starting to head south to look for garbage dumps or hook up with the grizzlies.

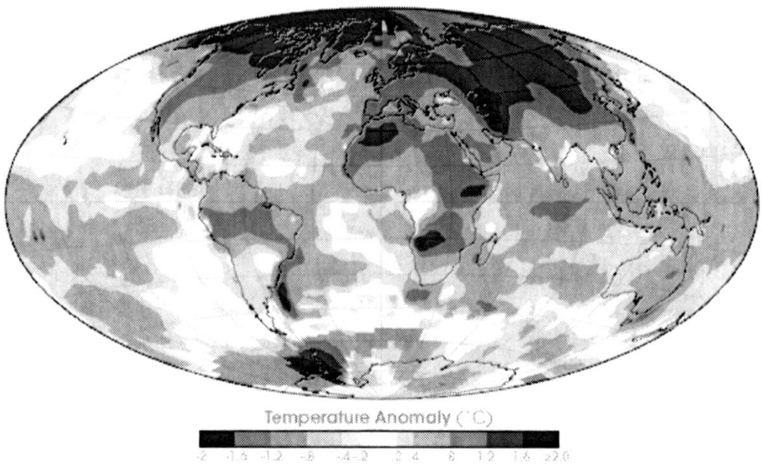

Fig. 95. World hot spots, obtained by comparing global temperatures in 2006 with averages from 1951-1980. Black means no data. Source: NASA GISS Surface Temperature Analysis.

A brief history of global warming

There had to have been global warming or places like Chicago still would be under a mile of ice, so warming is not all bad. Dickens' "*A Christmas Carol*" was published in 1883 and features some miserably cold weather. It was near the end of what now is called the "Little Ice Age."

Arguing away global warming also requires arguing away continental glaciation, so is there any evidence or is it all a hoax being perpetrated on people who had trouble getting past the fourth grade? Multiple working hypotheses were discussed in the last chapter; is continental glaciation the best and only explanation for the evidence?

1. *Far-fetched hard evidence.* Granite boulders such as shown in Fig. 96 are hundreds of miles from their homes in Canada, where they were part of a very old ground-off mountain range called a "shield area." Alternatives: Wind, no; running water, no; dragged along in glacial ice, yes.

Fig. 96. Granite boulders dragged into the U.S. by glaciers from Canada often have flat sides and skid marks. Handy and Spangler, Geotechnical Engineering © 2007 McGraw-Hill.

2. *Flat sides.* Boulders often have flat sides and grooves indicative of having been dragged along in ice. Other grooves also are etched into bedrock exposed in quarries. Glaciers, yes; icebergs? Icebergs come from glaciers and require water to float.

3. *Large lakes.* Basins for the Great Lakes appear to have been scooped out by glaciers, or else Paul Bunyan dragged his feet.

4. *Small lakes.* Lakes and bogs lay scattered across glaciated areas where there has not been enough time for them to be invaded and drained by fingers of erosion.

5. *Ice damming.* Lake Agassiz is mostly dry now. A lake bed that is over five times the size of North Dakota was created when a glacier dammed a major river that was flowing north at the time and sent it south to become the Missouri River. Remnants of Lake Agassiz are Lake Winnipeg, Lake Manitoba and Lake of the Woods.

It has been speculated that drainage of Lake Agassiz probably was rapid, possibly even catastrophic, following breaching of the ice dam, and the water may have flowed north into Arctic seas, possibly effecting some temporary changes in climate.

6. *Glacier boundaries.* The Missouri and Ohio Rivers approximately mark the southern extent of glaciation in the U.S.

7. *Soils.* Glacial deposits or "moraines" are mixtures of boulders, gravel, sand, silt and clay. Wind, no; water, the deposits would be separated into layers, no; slow release from melting ice, yes.

8. *Weathering markers.* Weathered clay layers separate and help to date deposits from different glacial advances.

9. *Glacial retreats.* Glaciers do not retreat because they do not back up. They keep flowing forward but the *edge* of a glacier will retreat if the rate of ice formation does not keep up with the rate of melting. A rim of hills is deposited when the two rates are approximately equal, the outermost rim is a "terminal moraine."

10. *Plants and animals.* Pollen preserved in lakebeds indicates that glaciation compressed climatic zones to the

south. Large animals got larger, the better to keep warm. Neanderthal Man wore animal skins to keep warm but since he did not have the sewing needle he had to walk bent over to keep his clothes from falling off.

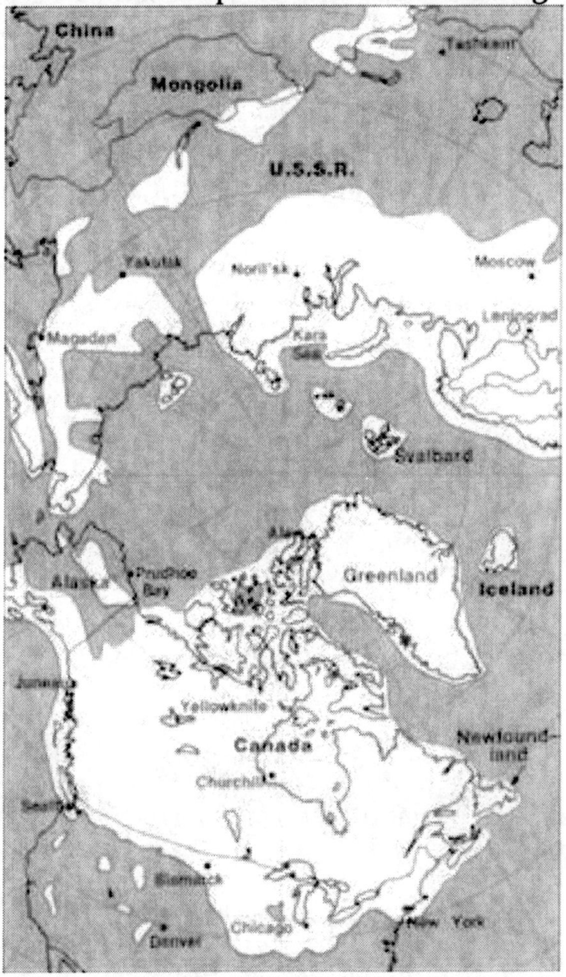

Fig. 97. Maximum extent of continental glaciers. U.S. Geological Survey.

Since Neanderthal had to keep a hand on his wardrobe, that may be why he never invented the bow and arrow, which takes two hands. The last known surviving

Neanderthal was at Gibraltar. He probably got pushed off.

11. *Ups and downs of sea level.* The name *delta* refers to a triangular shaped river deposit with the apex pointed upstream. Borings reveal that deltas also are deep, and that all major rivers emptying into seas cut canyons about 400 ft (120 m) deep, so either sea level was lower or rivers flowed uphill. The amount of sea level lowering is an indication of the amount of water that must have been locked away as glacial ice.

As ice melted and sea level came back up, river valleys are filled with sediment from melting glaciers or else remain partially filled as estuaries and fiords filled with cruise ships.

> *Some heavy arithmetic.* Oceans cover about three-fourths of the earth's surface and glaciers were on the other fourth, so the average ice thickness had to be at least 4 x 400 = 1600 ft. If only about a third of the land area was covered by ice, 3 x 1600 = 4800 ft. The ice mass had to be thicker in the middle or it would not flow. Since the height of a cone is three times the average, 3 x 4800 = 14,000 ft, or 2.7 miles. Add 10 percent because water expands with it freezes, 3 miles. Mile-thick ice that has been measured in Greenland, and thicknesses up to 2.8 miles exist in Antarctica.

12. *Sagging of the earth's crust.* In Fig. 97 Hudson Bay is close to the center of glaciation in North America, and tilted beach lines indicate that the area is rising. The ice was so massive that it apparently pushed down the earth's crust and it still is recovering.

13. *Wind blew and silt flew.* Rivers draining glaciers are overloaded with sediment and have intersecting braided channels and sand bars that were sources for dust. Dust piled up thickest on neighboring hills when winds blew nearly parallel to the rivers. The dust deposits are "loess," which is German for loose but is nothing personal. Loess soil is so loose that it does not hold a groundwater table, which enables it to stand in high steep banks. Radiocarbon age determinations indicate that loess in the U.S. and in Europe was mostly deposited during the last glaciation. Current loess deposition from a glacial meltwater river in Alaska is shown in a photo at the end of this chapter.

Fig. 98. Thick deposit of wind-blown loess silt border valleys of the Missouri and Mississippi Rivers. At the center left the author is at the end of his rope. Handy and Spangler, Geotechnical Engineering © 2007 McGraw-Hill.

Repeated glaciations

Cores of ice cut from glaciers in Greenland and Antarctica have layers like tree rings, and have checkpoints for dating such as volcanic ash from historical eruptions. At least equally significant, water temperatures that existed at the time the water evaporated are preserved by a ratio of two *isotopes* of oxygen. That is really cool. It is not the same as radioactive dating, which goes according to first-order rate equations. Instead, the evaporation rate of water partly depends on its molecular weight. Heavy molecules are lazy. It figures.

The H2 in H2O is hydrogen, atomic weight 1.0. The O is oxygen, atomic weight either 16 or 18 depending on the number of neutrons. The weight ratio of oxygen isotopes, O^{18}/O^{16}, flags the temperature of the water *at the time when it evaporated,* which in turn represents an average temperature of the seas. This is a huge discovery. It is like a surveillance camera with very long memory. It goes back many hundreds of thousands of years.

The Russian connection. The most famous ice core is the "Vostok," which was obtained by Russian Antarctic Expeditions during 1970-74 and 1982-83. The total length of core is 3,700 m (12, 140 ft; 2.3 miles). Ice samples were crushed and gases that were released were saved for analysis. The result is a continuous record of the thermal ups-and-downs during periods of glaciation, plus a record of air quality from air that has remained trapped in the ice ever since the snow fell.

The temperature record indicate four major glacial advances and meltdowns. This agrees with the count obtained from observations of soils. The data go back 420,000 years.

Underneath the ice sheet is a lake. It is Lake Vostok, and is creating some excitement because of all it may have held in cold storage for the last 420,000 years.

Peaks. The peak-to-peak changes in ocean temperature in Fig. 99 are about 10° C (18° F). That is more than the variation in the *Niños,* but may be less than what might be expected considering the consequences. A new glacial advance started about every 100,000 years. The same pattern is found in other ice cores from Antarctica and from Greenland.

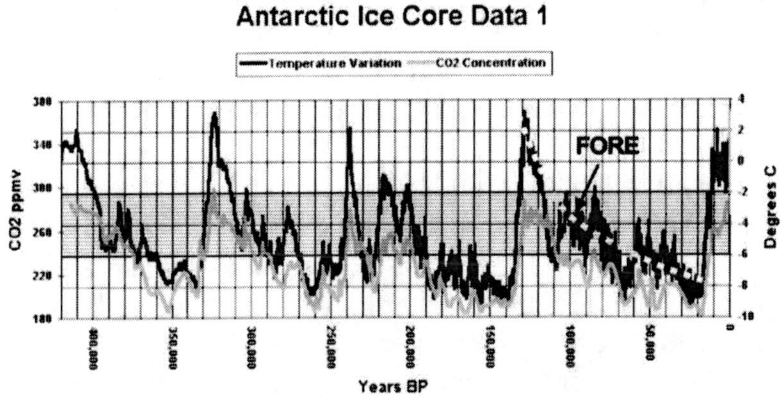

Fig. 99. Ice core data from Vostok Station, Antarctica, reported by Petit et al. in Nature, 399, 429-436 (3 June 1999). The shaded area shows the range in freezing temperatures of salt water, depending on the amount of salt. The darker line shows temperatures. Modified from a graph by Karl Davies, http://www.daviesand.com.© 1999.

The Milankovic interval. In 1930 a Serbian civil engineer, Milutin Milankovic, calculated a 100,000 year interval between major glacial advances, based on changes in the intensity of solar energy coming to the earth because of some wobbles, tilts and other imperfections in the way that the earth goes around the sun. The different factors act independently so sometimes they overlap and reinforce each other.

Milankovic's hypothesis did not have many takers until the ice core data came in. That about cinched it.

What about FORE?

In Fig. 99, each warming cycle was fast, indicated by the steep leading sides of the saw tooth pattern. Each cooling cycle was about 10 times slower so something was different: FORE only appears to work for the cooling cycles, shown by the dashed white line at the right in the figure, whereas warming was precipitous if one can call a thousand years precipitous. It hints at some kind of *positive feedback* that worked only for warming. That could be significant if the same positive feedback is at work today.

Positive feedback is like scratching an itch so it only creates more itching. Feedback can be a smile that leads to more smiles, or it can be a loud speaker that screeches because it can't stand to hear itself. What feedback could work for warming coincident with melting back of the glaciers but not for cooling and a re-advance?

Carbon dioxide already? Air trapped in glacial ice contains carbon dioxide that, as shown in Fig. 99, closely parallels the changes in temperature. This is heavy, because carbon dioxide is a "greenhouse gas."

A greenhouse gas lets the sunshine in, but after it converts into heat it can't get back out. Conversion increases the wavelength and decreases penetration. It is the really short wavelengths that give X rays good penetration. Light that comes in through glass in a greenhouse or solar panel, turns into heat and can't get back out. If a solar panel had only a mirror it would not work.

The only problem with carbon dioxide being the feedback for warming is where did it come from? *Neanderthals cooking marshmallows?* If the primary control mechanism was heat from the sun, what did that have to do with carbon dioxide when more green plants mean less carbon dioxide? It's time for some hypotheses.

One hypothesis has to do with carbon dioxide emissions from volcanoes, but they are unlikely to go off n schedule every 100,000 years. Another suggestion involve release of some of the huge amounts carbon dioxide that are held locked away in limestone, but huge amounts of energy are required to get them out to make lime and Portland cement.

Observations while duck hunting

Duck hunters get up before dawn, dress invisibly and paddle their invisible boats out into icy marshes where they toss out some duck decoys, break out the coffee and hard-boiled eggs, pet the dog and tell stories. After the coffee takes hold, a hunter may stand up and direct his attention at the water, when he may observe strings of tiny bubble rising through crystal-clear water. It looks like this would be a good place for fish but there isn't even a minnow. There has to be a reason for that.

What about those tiny bubbles? They presumably could be oxygen released by plants, but many places where the bubbles are common the plants are mostly dead and in the muck at the bottom of the lake. That opens a possibility that the bubbles are *methane.*

Methane is a rogue half-brother of carbon dioxide; whereas the carbon dioxide molecule is CO_2, one carbon plus two oxygen atoms, methane, CH4, is a carbon plus four hydrogen atoms. Methane is created by decay of organic matter where there is a shortage of oxygen, as in a marsh or landfill. Enough methane is produced from garbage in a large landfills to run electric generators. To top it off, methane is a greenhouse gas that is many times more effective than carbon dioxide. In air it eventually combines with oxygen to make carbon dioxide and water.

Might the carbon dioxide peaks in Fig. 96 actually represent methane peaks? Measurable amounts of methane were found in the Antarctic ice cores and show similar spikes, Fig. 103.

Circumstantial evidence. Fig. 97 shows a sudden dip in the methane content of ice cores that occurred about 12,000 years ago. This coincides with a temporary decline in global temperature called the *Younger Dryas,* named for an alpine flower. The timing also approximately coincides with the appearance of the Clovis spear point, and with draining of glacial Lake Agassiz.

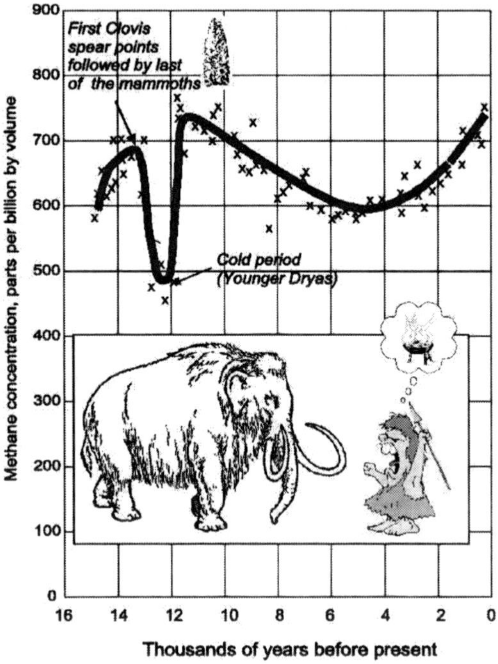

Fig. 100. Abrupt decrease in methane about 12,000 years ago coinciding with the "Younger Dryas," a severe cold period. Adapted from Felisa A. Smith et al., "Methane emissions from extinct megafauna," Nature Geoscience, 23 May 2010.

Is there a Clovis connection? One speculation is that appearance of the superbly crafted Clovis spear point and people who knew how to make and use it spelled doom for the mammoths, and for other large bests that may have been methane generators. Modern elephants trumpet from both ends, and one reason for their long noses is so they can reach up and get a little fresh air. After the megafauna were gone, something else, maybe herds of bison, must have filled in.

Prairie potholes. Swamps, marshes, peat bogs, landfills, sewage lagoons, cows and pulling grandpa's finger, plus a myriad of shallow lakes and peat bogs was left by the glaciers, are today's sources of methane.

Fig. 101. Poorly drained glaciated areas are likely sources for methane. Most have been drained for agriculture. http://landsat.gsfc.nasa.gov/

Hog confinements. Pork production has been industrialized, and from the point of view of the pig prospects are dim. A typical confinement operation will have a couple thousand hogs under one roof, and typically there is more than one roof. Downwind there is no need to speculate on nature of the aroma.

A hog can dump a gallon of excrement per day without giving it a second thought: 2000 hogs, 2000 gallons per day. It falls through slats in a steel grid the floor and is channeled into holding areas that emit some powerful odors. The gases are sufficiently toxic that they are dangerous, and workers entering a pit without an air supply may not get out alive.

The confinements usually are located away from towns, but there still can be problems that require serious lobbying of state legislatures. One possibility might be to collect the methane and flare it off, but another view is that this isn't anything that can't be solved with a little air freshener. Ironically, cows have been getting most of the blame when what they need is some better PR.

Feedback and sea level. In 1956 two oceanographers, Maurice Ewing and William Donne developed a hypothesis that may link the rise and fall of sea level to glacial cycling. They suggested that as glaciers melted and sea level rose, warm Pacific water would start to spill across the Bering Strait to warm and melt the ice off of Arctic seas. The open water in turn would evaporate water to make snow that would contribute to a return of glaciation that would lower sea level and allow the seas to again freeze over. A relation of snowfall to open water is shown in Fig. 102. Upper Michigan gets so much snow that houses have second-story doors.

One question concerning the Ewing-Donne hypothesis is whether the feedback would be symmetrical since the peaks in Fig. 99 are not symmetrical. Nevertheless sea level might be some sort of trigger for further action.

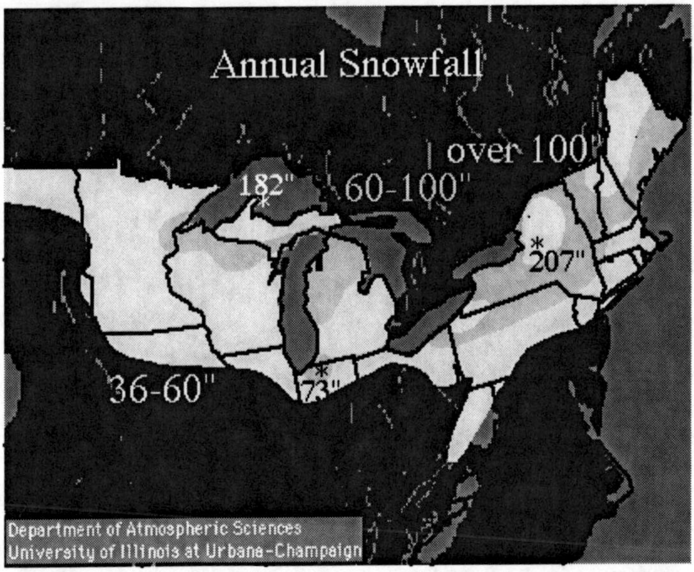

Fig. 102. Lake-effect snow can pile up higher than an elephant's eye in a single season. Dept. of Atmospheric Sciences, Univ. of Illinois.

Permafrost feedback

Arctic permafrost, or what used to be considered permanently frozen ground, is rapidly melting and according to data from Siberia is releasing surprising amounts of methane. Methane in permafrost is contained in ice nodules called clathrate, which is the ice that burns. About 30 percent of the area of Russia is in permafrost.

Fig. 103 may help to explain the methane release, as a large swirl of organic soil has been pushed deep into the permanently frozen layer by lateral squeezing from vertical ice wedges. The wedges form when water seeps down into open shrinkage cracks that re-open when the soil shrinks every winter.

Fig. 103. Organic soil folded deep into Alaska permafrost soil by lateral pressure from expanding ice wedges. The exposed wedges have melted out from both sides. From Handy and Spangler, Geotechnical Engineering ,© 2007 McGraw-Hill.

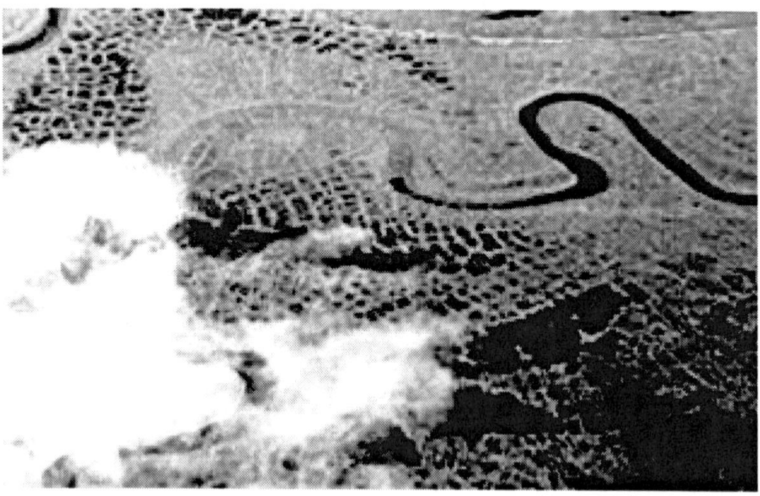

Fig. 104. Melting of permafrost on the north slope of Alaska. The polygonal pattern is created by ice wedges. From Handy and Spangler, Geotechnical Engineering © 2007 McGraw-Hill.

Carbon dioxide today

Fig. 105 shows carbon dioxide contents measured in air at Hawaii, which is a long way from any major coal-burning enterprises. Today's carbon dioxide content is about 35 percent higher than the range measured during the Pleistocene ice ages, and so far gives no hint of leveling off. This is new territory.

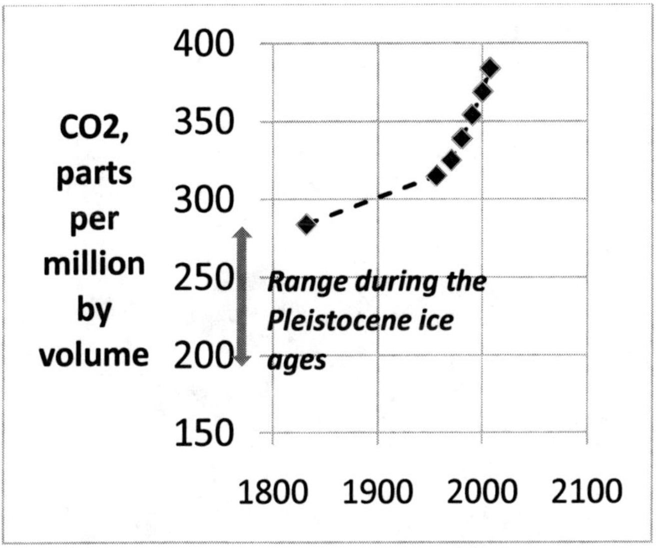

Fig. 105. Carbon dioxide contents measured in air at Mauna Loa, Hawaii. Data are from NOAA, the National Oceanic and Atmospheric Administration.

The PETM maximum. An earlier episode of warming is called the "*Paleocene-Eocene thermal maximum,*" about 55 million years ago. Global temperatures bumped up about 5° C (9° F) and the cause is not known. One speculation is that volcanism may have caused a methane burp from the oceans, where considerable amounts of methane resides in deep water as clathrates. Whatever the cause, the effect was sufficient to allow palm trees to grow in Alaska. The top of the food chain was occupied by a cat-like carnivore called a *creodont,* and the only primates looked like big-eyed squirrels. A fossil snake 43 ft (13.1 m) long was recently was discovered in a coal mine in Columbia, and had enough room in its gut for a freight elevator. PETM is not to be confused with PETA, which probably would have taken up the cause of the snake.

The Little Ice Age

Shown in Fig. 106, some serious cold weather occurred between about 1600 and 1900, after the Medieval Warm Period got everybody's spirits up. During the Little Ice Age crops failed, it snowed in summer, and people became

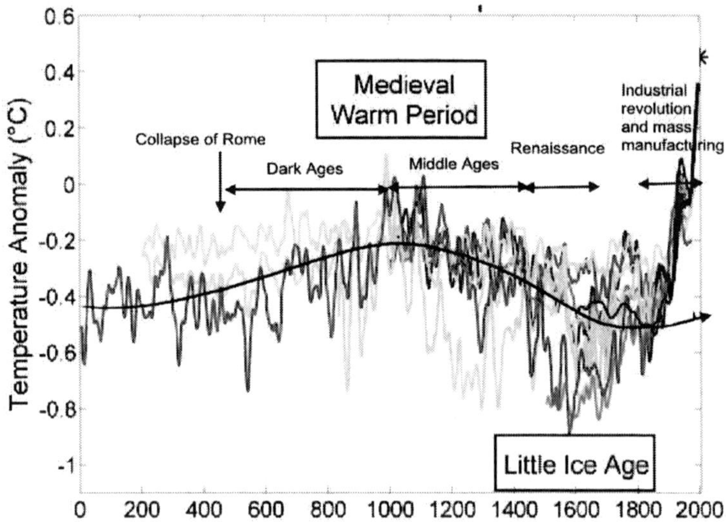

Fig. 106. Global temperatures during the last 2000 years from 10 different sources.. All agree that a sudden jump began in the late 1800's, coincident with start of the Industrial revolution. Thus ended the Little Ice Age. Modified from http://www.globalwarmingart.com /wiki/image:2000_Year_Temperature_Comparison_png.

highly agitated. In 1793 in France they were so agitated that they removed the heads of state, one at a time. As previously mentioned, in 1818 Mary Shelley caught the mood with her book "*Frankenstein,*" and Dickens' "*A Christmas Carol*" was a testimony to the exorbitant price of coal. The word "blizzard" blew in from Iowa during the winter of 1880-81.

The arrow in Fig. 106 shows what might have happened were it <u>not</u> for global warming: We probably still would be suffering the Little Ice Age. As late as 60 years ago a major worry of the professional worriers was that glaciation might return with all of its inconveniences.

Fig. 107 focuses on recent temperature trends, which are different in the two hemispheres and close to the equator. Temperatures in the northern latitudes are the most variable. Coincidentally that is where the pollution is, and the peak-to-peak trend is the same as the more consistent changes farther south. They indicate a warming of about 1° C (1.8° F) every 100 years.

The upper graph in the figure shows that air pollution is not the only game in town, as between 1940 and 1970 temperatures in the northern latitudes did not go up; they went down. Then they did a catch-up and went up again. Critics of warming point out that temperatures also have leveled off in the last 15 years, but let's not get too excited; it has happened before. Why did it happen? Remember that the rate of warming appears to involve positive feedback whereas the rate of cooling does not.

Coal trains

Burning high-sulfur coal created an international incident when acid rain started killing pine trees, which led to the increased use in the U.S. of low-sulfur strip-mined in Wyoming. A typical coal "unit train" is shown in Fig. 108.

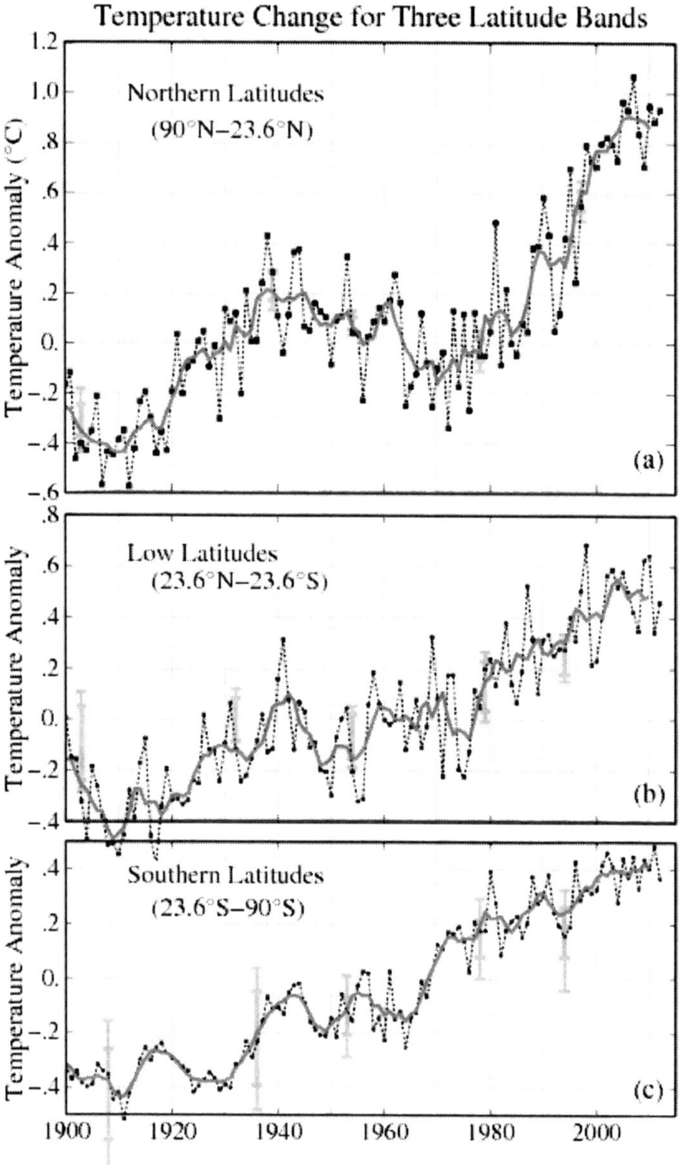

Fig. 107. Recent changes in global temperatures in northern and southern latitudes and in a band along the equator. http://data.giss.nasa.gov./gisgraphs_v3/.

A coal unit train rumbles out of Wyoming about every 20-30 minutes, 24/7. Each railroad car carries about 100 tons of coal, and there are at over 100 cars in a train. That comes to 100 x 100 = 10,000 tons of coal per train, with 50 or more trains making the trek every day.

A ton of coal makes about 3 1/2 tons of carbon dioxide that is dumped into a carbon dioxide cycle, green plants taking it in and releasing oxygen, and animals breathing in oxygen and exhaling carbon dioxide. The effect of the additional CO_2 therefore is not so monstrous as it might be. More carbon dioxide comes out of tail pipes of automobiles, and an estimate by the U.S. Dept. of Energy that six trillion tons of carbon dioxide are being released from burning fossil fuels in the U.S. every year.

Fig. 108. Coal unit train rolling east to generate electric power. Every ton of coal burned makes about three and one-half tons of carbon dioxide, over 3500 tons per train that rolls by at a rate of three per hour. Author photo.

Wyoming coal represents about half of the coal burned in the U.S. but it still is less than 15 percent of the coal burned in the world. China holds the lead, and the air there pollution is health-shattering and abysmal, causing a measurable reduction in life expectancy for people who don't even smoke.

The good news/bad news is that the carbon dioxide does not stay in the air; it comes down in rain and ends up in the oceans.

Not so fine. Acidity is measured in terms of "pH," pronounced with a pee and an "h." Carbon dioxide dissolves in water, where it makes a weak acid that puts the tang in soda. The acid still is strong enough to dissolve limestone and create caverns.

The pH of water in the oceans has lowered about 0.1 since the mid 1800s, but pH is on a logarithmic scale. Brought back into the real world that can mean an actual increase in acidity of about 25 percent, and coral reefs are objecting. They are getting out of the mood to reproduce, which could eventually doom a lot of marine species. If this continues it may not be possible to order oysters on the half-shell because they won't have an shells. It could be enough to turn a sea cucumber into a pickle.

Up with sea level!

Sea level is risen! It rose about 400 ft (120 m) when continental glaciers melted, and glacial ice is still melting. Floating ice masses such as pack ice on Arctic seas and shelf ice in Antarctica don't count; ice cubes melting in a glass of water don't raise the level of the water.

According to Fig. 109 during the last 12,000 years, which coincides with melting of the glaciers, the rate of sea level rise has been proceeding according to FORE. That is reasonable, as the rate should slow down as the world runs out of ice. The ultimate projected additional rise is 1.6 m (5.2 ft), which still leaves Antarctica. Other estimates give as much as 2 m (6.6 ft) but they are complicated by the many different factors involved. According to FORE the current rate should be about 0.54 mm per year, or 5.4

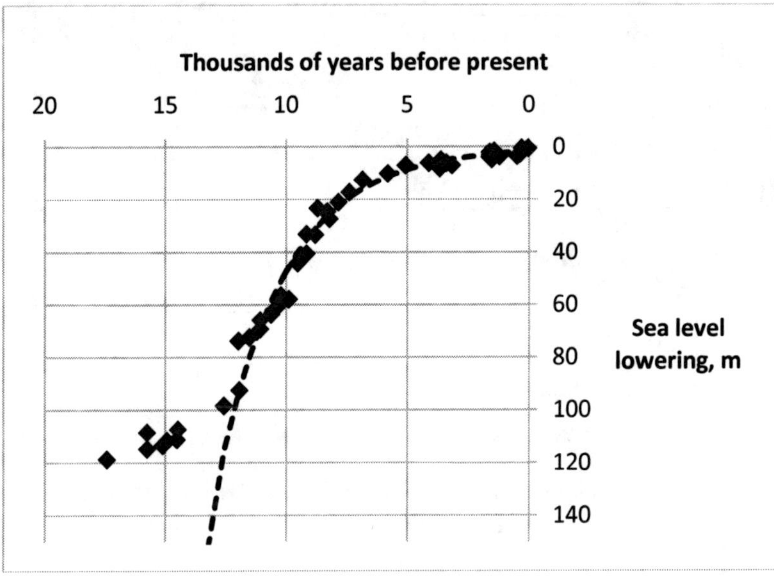

Fig. 109. Sea level rise has followed FORE ever since the last of the glaciers started some serious melting about 12,000 years ago. Data are from R. G. Fairbanks (1989), "A 17,000-year glacio-eustatic sea level record..." Nature, 342:637-642.

cm (2.1 inches) in 100 years. Also, even that rate should be slowing down. The bad news from satellite measurements, Fig. 110, is that the rate is over 5 times as fast, 3.1 mm per year. That being the case, the next 100 years could see a rise in sea level of more like a foot. Fortunately the last few years indicate some slowing down--but temporary slowdowns have happened before. If the rapid increase continues, may it signal a *major event?* Remember what happened with the pole vault.

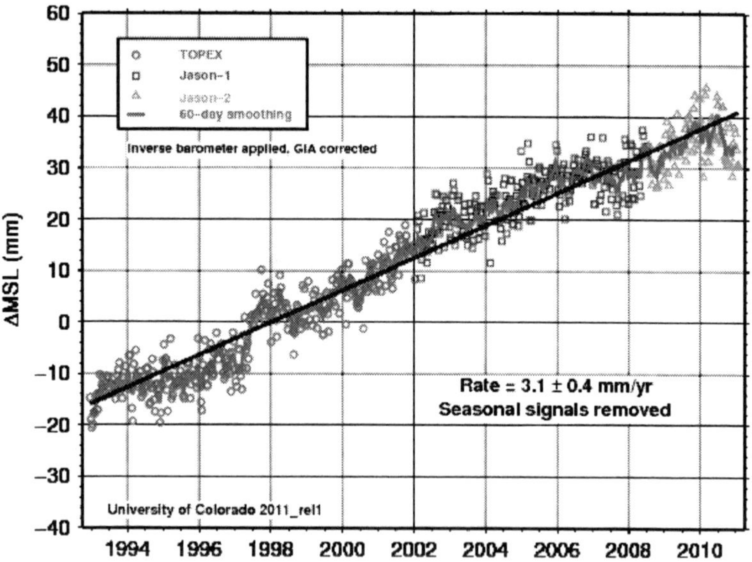

Fig. 110. Sea level rise that is five times faster than indicated by FORE.. Sea Level Research Group, Univ. of Colorado.

Causes for the speed-up. Some possibilities: (1) Carbon dioxide production is concentrated in the northern hemisphere and is being released at an unprecedented rate into the atmosphere, Fig. 105. (2) In northern latitudes the rate of warming rapidly increased between 1970 and 2000, Fig. 107, so maybe now it will let up a little. (3) Methane production from lakes, animals and melting of permafrost is concentrated in the northern hemisphere and has increased three-fold in the last century, Fig. 103. (4) Some other reason or reasons; it's hard to tell.

Corrective measures

150 years ago a little additional global warming sounded like a good idea, but let's not get carried away. Some undesirable consequences that already may be working

against us are more severe weather patterns, an additional rise in sea level, and eventually a demand for more air conditioning, all of which can create troubling consequences.

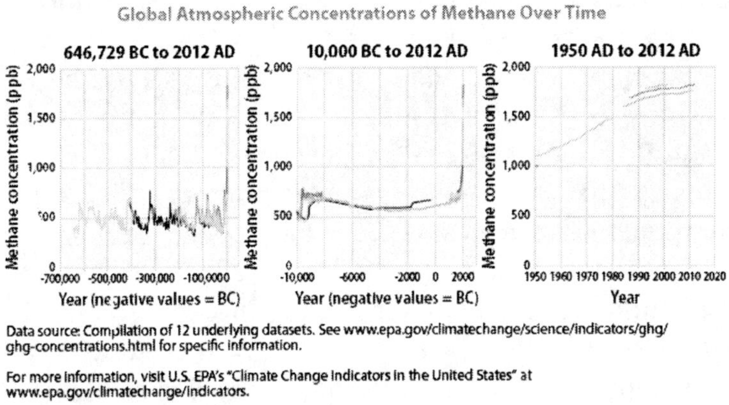

Fig 111. Methane contents of the atmosphere determined from Antarctic ice cores. U.S. EPS.

So what, if anything, can be done about it? There is been no lack of ideas, just a lack of enthusiasm,

Burning natural gas instead of coal will reduce carbon dioxide emissions by half and also save money, which is not good news for coal mining. Every industry has its day in the sun.

The number of miles per gallon of automobiles has doubled, partly as a result of federal mandates, reducing emissions and saving money. There is little opposition as long as pickup trucks are exempted.

More cars are running on electricity, which improves efficiency and ultimately plugs into power plants. Even better would be to park the car under a windmill.

Beyond that the choices are difficult, a simple procedure being to deny everything and put one over on the next generation.

Here are some more ideas: (1) Extract carbon dioxide from power plant smoke and pump it into the ground where it will stay forever. This would attach an expensive process to the cheapest, burning. We think getting rid of a few hundred tons of nuclear waste is hard; wait until we try this one, at the rate of over a trillion tons per year. (2) Invest in more windmills. Winds do not always blow at the most convenient times, but there are many ways to store and release power. (3) Turn on the nuclear, which is unlikely to get much popular support because a component of design appears to have been is to skimp and save money. (4) Throw tiny flakes of reflective plastic into the wind to reflect a portion of the sun's energy, and then get ready for the lawsuits. Any tinkering that affects everybody like it or not is asking for it. Tinkering has to have the appearance of being unintentional.

Working up a hypothesis

• Global warming had to happen or glaciers still would still be grinding on places like Cape Cod, Cincinnati, Chicago, Omaha and Saskatchewan.

• The march of the glaciers every 100,000 years appears to relate to wobbles and other imperfections in the way the earth handles itself.

• Each glacial cooling went according to FORE but subsequent warming got pushed along about ten times as fast, a screech caused by positive feedback.

• Methane and carbon dioxide contents in air took a serious jump every time there was warming, and methane in air turns into carbon dioxide. Methane is generated in abundant marshes and bogs created by melting glaciers and concentrated in the northern hemisphere.

• Large amounts of carbon dioxide have been released starting with the industrial revolution, and methane is being released from cattle feedlots and industrialized hog production.

• The good news is that a rate of increase in temperature that appears to be unprecedented may be what put an end to the Little Ice Age and may even have put off another episode of continental glaciation. Given that choice warming would be better.

• Sea level came up about 120 m (400 ft) during the last glacial meltdown. The rate of rise depends on the amount of ice left available and therefore has closely followed FORE.

• The current rate of sea level rise is many times faster than the forecast by FORE so there must be a change is some critical condition.

• The carbon dioxide content of air is about 30 percent higher than it was at any time during previous episodes of global warming.

• Arctic permafrost is rapidly melting and releasing encapsulated methane. This may be unique to the current episode of warming.

• More stormy and erratic weather patterns are consistent with warming.

• Denial still can be better if it sooths the soul.

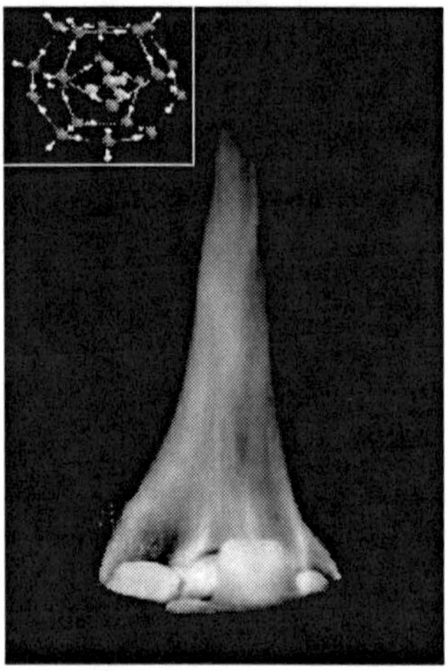

Fig. 112. Melting of permafrost means melting of contained methane clathrate, the ice that burns. It is also is abundant in deep cold waters of the seas. U.S. Geological Survey.

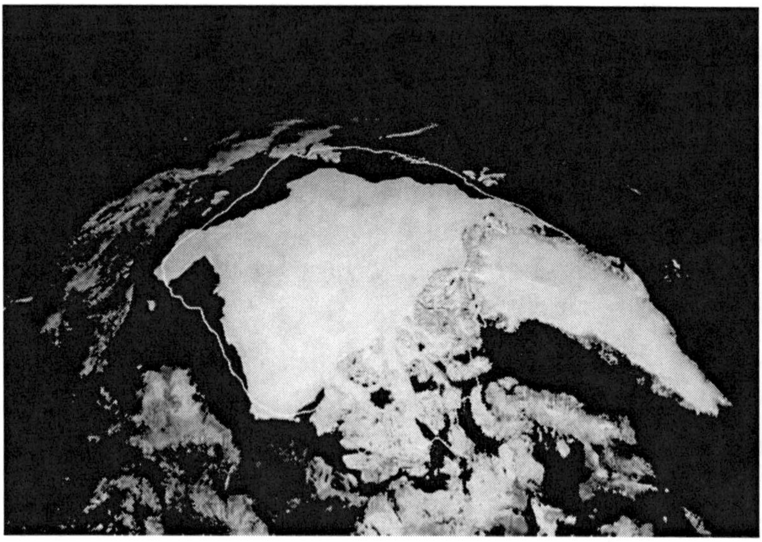

Fig. 113. Retreat of the polar ice cap and why the petroleum industry is getting excited. The bright line shows the thirty-year average summer minimum. NASA.

Fig. 114. Looking north from the north coast of Alaska during a maximum melt of pack ice in 1955. Author photo.

Fig. 115. Fifty years later, and polar bears are turning white with worry. ©David Thoreson Photography, Blue Water Studios, Okoboji, Iowa. Compare with Fig. 6-8 in Chapter 1.

APPENDIX
The rest of the best

The title of this book did not promise the future of *everything;* it said *practically* everything. The plan is to present a few examples and show how FORE works, then fill in enough blanks so anybody can do it. All that is required is curiosity, patience, wakefulness, and a computer with a spreadsheet.

Readers whose constitutions do not tolerate math should not read past the next paragraph as it may make you sick.

Basic assumption. The basic assumption is that a rate of change depends on the amount of departure from some final equilibrium state. That means that a process has to stop when some key ingredient is all used up. That is not complicated.

A rate of change can be expressed by *dy/dt,* where *dy* represents a tiny bit of change and *dt* is a tiny snatch of time. The / means to divide, as in miles/hour. The assumption therefore is:

$$dy/dt = k \cdot y \qquad \text{(A1)}$$

That is a definition. It means that a rate of change *dy/dt* at any particular moment equals some number means *k* times another number *y*. The *k* is a *constant* because it does not change unless we ask it to. The • means to multiply so x can be saved for other matters. Sometimes it is left out:

$$dy/dt = ky$$

Now comes the *algebra.* After all these years, a use for algebra, and but it gets easier the second time around.

First clean up the left side of the equation by multiplying both sides by *dt:*

$$dy = ky\ (dt) \qquad \text{(A2)}$$

The parentheses are put around *dt* so everybody will understand it is one number and not *d* times *t.* Then since everybody knows that we can eliminate the parentheses.

Now, for a reason that will be shown in the immediate future, divide both sides by y:

$$dy/y = k\ dt \qquad \text{(A3)}$$

Now comes the karma. It is called *calculus.* It was invented by Isaac Newton around 1665 before the mercury vapors got him confused.

Integral calculus uses a crooked ∫ and sums up the *d's.* According to nuances of integral calculus,

$$\int(dy/y) = k \int dt \qquad \text{(A4)}$$

The *k* slips out from under the ∫ because it is constant, and ∫dt sums up all the little *dt's* to equal a total time represented by *t.*

The integration of dy/y is tricky and is a *logarithm.* It is not just an ordinary logarithm; it is a *natural* logarithm, which makes it more expensive.

After integration the equation becomes

$$\ln y = k\ t + b \qquad \text{(A5)}$$

where "ln" means natural logarithm. The *b* is called a "constant of integration." The proof for this is beyond the scope of this book and certainly beyond the comprehension of the author.

The base for a logarithm is 10, and the base for a natural logarithm is *e* = *2.71828,* which does not seem very natural. Most people count with discrete numbers of fingers unless they have a razor up their nose. Fortunately a natural logarithm can be converted into a real one by multiplying by another constant, which can be tossed in with the first constant to make a super constant, but that's how it is with constants. Then

$$\log y = K \cdot t + B \tag{A6}$$

It took a while to get there, but that is the equation for a straight line! It even looks a little like $y = ax + b$ except that y *is* a logarithm. In order to look a little more savvy the equation can be written

$$\log y = at + b \tag{A7}$$

This simple equation says the following:

If logarithms of numbers represented by y are plotted on a graph with corresponding values of time t, the result should b a straight line.

If the line is not straight, FORE does not work.

Equation (A7) has a limitation but since it is *analytical* equation it can get an *analytical* fix. It's like par golf and going for the next hole.

The y value represents some rate of change, and the limitation is that it may not go all of the way to zero. The equation comes from chemistry where reactions end at zero. Physical processes may not end at zero; sometimes after they slow down they keep going. That can be taken care of with one more letter in the equation. We call it c:

$$\log(y - c) = at + b \tag{A8}$$

That is the super-duper FORE equation.

One more observation: If a is positive it means the y is increasing, and if a is negative, y is decreasing. However, $(y - c)$ can't be negative because there is no logarithm for a negative number. The fix is easy: Make it positive. Would that tax forms would be that easy.

The code for making sure that a number is positive only involves straightening out the parentheses. Then

$$\log |y - c| = a \cdot t + b \tag{A9}$$

and that's all there is to it.

Testing FORE

Equation (A9) has an extra unknown quantity, which can play hell with a relationship. It has to do with degrees of freedom and this one appears to have too much freedom. However, there is another important bit of information that allows the equation to be solved: There should be only one value for *c* that makes a straight line. All that is required is to try out different values for *c* and pick the value that makes the best straight line. If there is no value that makes a straight line, it means that FORE does not work.

Major events. Quite often a graph of the data has two distinct curves that are separated by what in the language of FORE is a *major event.* For example, changing the kind of pole used in the pole vault was a major event because it set an entirely new course for setting new records. In that case all that is required is separate the data and find two vales for *c.*

Translating records. The two values for *c* before and after a major event represent ultimate values where a system is at equilibrium and there is no further change. The ratio of the two *c* values then can be used to translate between old and new records and asses the significance of the change. For example, changing the kind of pole used in the pole vault increased heights by 23 percent.

The same procedure can be used to translate women's sports records into male equivalents, thereby putting numbers on a *gender factor.*

The R^2 *test.* A statistic called *R-squared* is a measure of how well data lay along a straight line, and therefore is useful to rapidly measure effects from changing *c.* The test procedure involves changing *c* while keeping an eye on R^2 and selecting the one that gives a maximum. For example, in order to find a maximum value and the best value for *c*. This method of solution is called *trial-and-error.*

The setup. A computer spreadsheet has lines and columns. The columns can be labeled across the top of the spreadsheet as follows:

• Column A: Time. (This is *t* in the equation.) This does not necessarily have to be time; for example it could represent different distances. It is called the *independent variable.* Numbers in Column A normally will increase incrementally to reflect how the data were obtained. In the example the time increment is 10 years starting with the A4 cell with 1920. A5 is written as A1+10, and this formula is copied and pasted down the rest of the A column.

• Column B: These are the measured values and are entered individually. These are *y* in the equation.

At this time an xy graph should be drawn by swiping the two columns, with Column A values plotted in the x direction and Column B values in the y direction. The graph then should be inspected for sudden changes, but this also can be accomplished later.

• Column C: Trial ultimate value, c in the FORE equation. This trial value is entered in Cell C4 and copied for the full length of the column. The C4 value will be changed over and over again in order to get the highest R^2 *value in cell E4.*

• Column D: Write =log(y - c) in Cell D4 and copy for the length of the column.

• Column E: Write =RSQ(A1..A$_n$,D1..D$_n$) in cell E4, where "n" designates the total number of data entries.

• Column F: Theoretical values for y in the FORE equation. This calculation requires a graph of Column A numbers for x values and Column D numbers for *y* values. When the value in cell C4 is correct, this graph should make a straight line. The manual procedure is to change C4 and observe the effect on *R2*. If it goes up, the change was in the right direction; if it goes down, reverse the direction. Solution is rapid and can be performed with a *macro*.

After a best value for C4 has been obtained, right-click on any data point on the graph and click "Add trendline." Click the boxes for "Linear," "Display equation on chart," and "Show R^2..." Click "Close." The equation for the line and the R^2 will be written on the graph. The R^2 should be the same as in spreadsheet cell E4.

Calculating values for Column F. This uses the equation on the graph and has the form $y = ax + b$. The a and b values are picked off of the equation written on the graph. The expression for F4 is

=-C4+10^(-a*A4+b)

The equation is copied for the full length of the column. If answers are negative, change the sign so they are positive.

Extending Column F into the future. This is one of the most important uses for FORE. It is accomplished by extending Columns A, C, and F as far as desired. As changes become smaller, larger increments can be selected for Column A values.

Adding formula values to the first graph. Re-open the first graph and add another series with Column A values in the *x* direction and Column F values in the y direction. Remove the markers so they will not be confused with the actual data. The C column ultimate value should be written in a text box on the graph.

To anybody who has read this far: Hold your head high because now you are an official Nerd and qualify to have an opinion, but that can endanger your status because most nerds stay cool and pretend not to have opinions.

Example calculation: Cod catch in the North Atlantic

In the following table the C column entry already has been varied, with a value of 44 thousand tons giving the highest number in E1. That value of 0.9968 is close 10 1.0000, indicating that the data are precise and the analysis okay, and things do not look good for fishing.

1	A	B	C	D	E	F
2	Year	Thousand Tons	Ultimate	log difference	R^2	Formula value
4	1920	300	44	2.408240	.996803	287.2
5	1930	225	44	2.257679		229.1
6	1940	190	44	2.164353		185.0
7	1950	150	44	2.025306		151.3
8	1960	130	44	1.934498		125.7
9	1970	105	44	1.78533		106.2
10	1980	95	44	1.70757		99.4
11	1990	80	44	1.556303		80.0
12	2000		44			71.5
13	2010		44			64.9
14	2020		44			59.9
15	2030		44			56.1
16	2040		44			53.2
17	2050		44			51.0

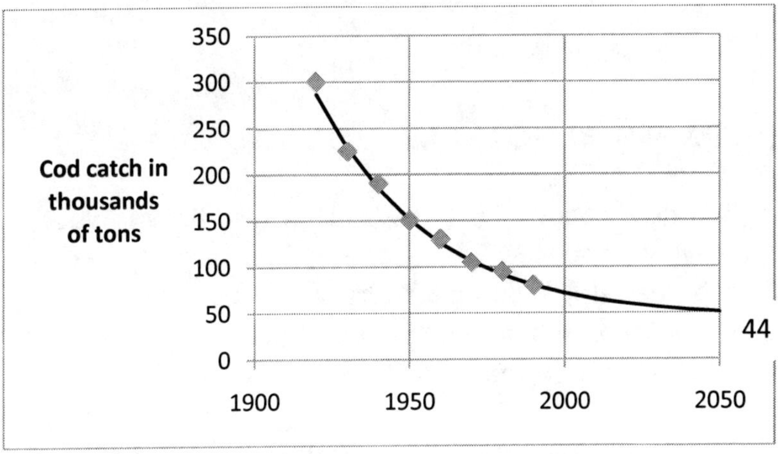

Fig. A-1. Dots show the data; the line is drawn by FORE.

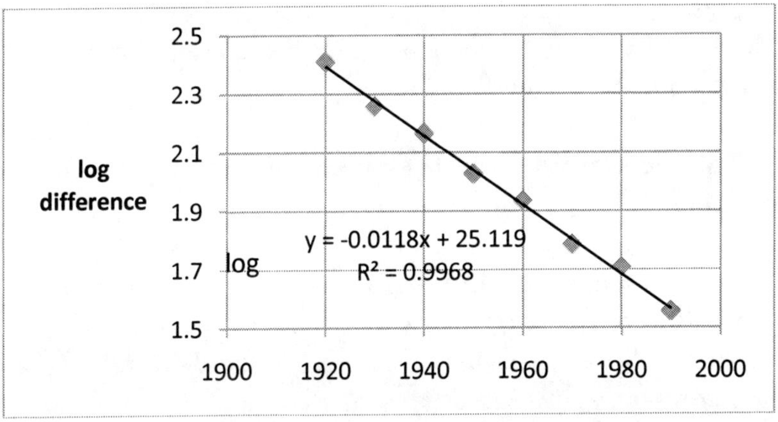

Fig. A-1. Best straight line and highest R^2 value obtained by varying c in Equation A-9, and the equation used to draw the FORE line in Fig.A-1. The projection is for c to eventually diminish to 44

JUST A FEW REFERENCES

Brandt, Anthony (2010). *The Man who ate his Boots.* Alfred A. Knopf, New York. Camilli, R., Bowen, A,. Reddy, G. M., Seewald, J. S., and Yoerger, D. R. (2012). "When scientific research and legal practice collide." *Science 337:1608-1609. Aftermath of the BP Gulf oil spill.*

Cellulosic ethanol: http://www.eia.gov/biofuels/workshop/pdf/paul_kamp.pdf

Dyson, Freeman (2012). "Is science mostly driven by ideas or by tools?" *Science 338:14 Dec./1426. Empirical research fosters new theories.*

The Economist (2012). "Backward ran sentences..." 9/1/2012. *Time goes only one way.*

Federal Railroad Administration (2009). "Comparative evaluation of rail and truck fuel efficiency on competitive corridors. *Rolling on rails takes less energy than rolling on highways, and taxes don't buy the rails.*

Handy, R. L. (1972). "Alluvial cutoff dating from subsequent growth of a meander." *Geol. Soc. Amer. Bull.* 83:475-480. *First use of FORE for a physical process.*

Handy, R. L. (1973). "The igloo and natural bridge as ultimate structures." *Arctic,* I26:4:276-281. *Darwinian architecture: Survival of the fittest.*

Handy, R. L. (2002). "First-order rate equations in geotechnical engineering." *Amer. Soc. Civil Engr. Jour. of Geote* Chapter *Engineering* 128:5:416-425. *Even including pole vaulting? You're kidding.*
http://www.atmos.washington.edu/gcg/RTN/rtnt.html. *El Nino and climate change.*

Keating, Bern (1970). *The Northwest Passage.* Rand McNally & Company, Chicago. *A publicity piece with photos of the 1969 Manhattan expedition, made when ice was thinnest. The 1970 expedition was made when ice was thickest.*

Lear, W. H., and Parsons, L. S. (1993). "History and management of the fishery for northern cod in NAFO Divisions 2J, 3K and 3L." Can. Bull. Fish. Aquat. Sci. 226, pp. 55-89.

Lee, D. ., Herman, J. D., Elsworth, D., Kim, H. T., and Lee, H. S. "A Critical Evaluation of Unconventional Gas Recovery from the Marcellus Shale, Northeastern United States". *Energy Geotechnology*

Mayugai, M. N., and Allen, D. R. "Subsidence in the Wilmington Oil Field, Long Beach, Cal., U.S.A.
http://www.saveballona.org/gasoilfields/WilmSubGC.pdf. *Sinking from pumping.*

Robelius, Fredrik (2007). "Giant oil fields—the highway to oil." *Doctoral dissertation available online from Uppsala Univ., Sweden. Inevitable decline of new discoveries.*

Sports Illustrated (June 3, 1974). "Gold to Ivory, Ashes for Tony." SIVault network site. *Information on Ivory Crocket.*

Takeshi Horie et al., 4th International Crop Science Congress, Brisbane, Australia, 2004. *Rice production.*

Thomas, C. P., et al. (2009). "Alaska North Slope oil and gas a promising future or an area in decline?" U.S. Dept. of Energy DOE/NETL=2009/1385. *Skip the politics.*

Thoreson, David (2008). *An Arctic Journey in a Changing World.* Iowa Public Television, Des Moines. *(Recent sailing through the Northwest Passage on a sailboat.)*

Index

Afterword: Getting serious about humor

Some people take offense if a humor is permitted to prance around and trip over serious topics like science and mathematics. The author has received some reprimands, like "Some people take their profession seriously!" The answer to that is that humor can be dead serious—particularly if there is a comma between the "dead" and the "serious." Life is short, so why not have a little fun with it?

In *The Act of Creation* (The Macmillan Company, 1964), Arthur Koestler draws parallels between jokes, the arts, and inventions. He points out that all can involve sharp and unexpected turns that have a different meaning: When Henny Youngman said, "Take my wife," he did not really mean for somebody to take his wife, not until after he said "please." Modern art can inspire a smile or two but of course it must be done discreetly. Some inventions start out as jokes, which is better than starting out as inventions and ending up as jokes.

Humor characteristically involves walking a tight-rope, which means that it can't go anywhere without the rope. Comics often get their start by making fun of themselves, then pump up the risk by doing impressions of other people, and take it from there. Imagine a prisoner in solitary trying to entertain himself by performing impressions of a wall.

This book takes a jab at a critical topic of the day, which is science education. It uses a scientific approach to project data into the future, and there are no guarantees. Sorry if we offend. Another alternative is to just sit back, loosen up and enjoy.

—The Author

About the Author

Richard Lincoln Handy was born in Chariton, Iowa, but since that was a long time ago he has no recollection. He grew up in Des Moines, attended Roosevelt H.S. and then graduated from Clayton H.S. in

Missouri, and all he can remember is that it was very hot. Dick enrolled in Iowa State University before it was a University, and drifted from engineering to geology with a dash of soil science and bit of journalism. As luck would have it, he was in a class taught by Leonard Feinberg, an inter-nationally known scholar of humor and satire. They became close friends.

After obtaining a Ph.D. in 1956, Dick was appointed an Assistant Professor of Civil Engineering. As part of his duties he was named to various committees including a committee-committee that was set up to organize the committees, which may have been part of the transition from being a College to being a University. That was when Dick quit going to committee meetings, which gave him a lot of free time for haircuts and contemplation. He went up through the ranks, no pun intended, clicking into full Professor in 1983 as a consequence of putting on a little extra weight. In 1986 he was named a Distinguished Professor as a step toward early retirement.

Meanwhile Dick put out the usual drivel of technical papers and textbooks, and some extraneous drivel in a newsletter called "Screenings" and a book, *The Day the Houser Fell,* that slipped in some jokes. Dick retired from classroom teaching in 1991 but since he had accumulated a bunch of patents he has a company that manufactures and sell some of his inventions but they are highly specialized. He is a great-grandfather but but has no further plans.

CPSIA information can be obtained at www.ICGtesting.com
Printed in the USA
LVOW06s1335041213

363741LV00004B/19/P

9 781937 327316